# Mobile App Development Basics: Create Apps That Work

## Learn App-Building Fundamentals for Android and iOS

Isabella Grey

# Table of Contents

# INTRODUCTION

Building mobile applications for Android and iOS is made easier with the help of "Mobile App Development Basics: Create Apps That Work: Learn App-Building Fundamentals for Android and iOS," which offers a simple and useful framework. Whether you are a novice or an experienced developer, this book provides a methodical approach to comprehending the core ideas of mobile app development.

Mobile apps are now a necessary component of daily life due to the sharp increase in smartphone usage. Companies, entrepreneurs, and developers continuously develop applications to improve user experiences and expedite certain tasks. However, creating useful and intuitive software takes more than simply an idea. Technical know-how, design concepts, and an awareness of how various platforms function are all necessary.

This book covers all the important facets of developing mobile apps, from configuring the development environment to creating user-friendly interfaces and adding necessary features. In addition to learning about popular frameworks and best practices for developing responsive and captivating applications, readers will acquire insights into the distinctions between developing apps for iOS and Android. The chapters cover key subjects, including UI/UX design, backend integration, testing, debugging, and app deployment, and are organized to walk you through the full development process.

The book covers technical ideas and crucial factors, including security precautions, performance optimization, and methods for releasing apps on the Google Play Store and Apple App Store. You will have a firm grasp on creating a fully functional mobile application by the end of this book, along with the self-assurance to see your app ideas through to completion.

This book offers the fundamental information and helpful advice required to succeed in mobile app development, regardless of your goals: creating your first app, refining an existing project, or learning more about mobile technology.

# CHAPTER I

# Getting Started with Mobile App Development

## Understanding the Mobile App Ecosystem

The mobile app ecosystem, a large and ever-changing digital landscape, has completely transformed how people use technology. The lifecycle of mobile applications is shaped by a complex web of users, platforms, developers, marketplaces, and regulations. It is necessary to examine this ecosystem's essential elements —such as app development, distribution channels, monetization tactics, user engagement, security issues, and emerging trends. In a market that is becoming increasingly competitive, how these factors interact determines whether an app succeeds or fails.

App development, which includes designing, creating, testing, and distributing applications, is at the core of the mobile app ecosystem. Developers utilize a variety of platform-specific programming languages and frameworks, including Kotlin for Android, Swift for iOS, and cross-platform technologies like Flutter and React Native. Market research is usually the first step in the development process to determine user wants and competitive gaps, which generally takes an organized approach. Wireframing and UI/UX design come next, ensuring the app is aesthetically pleasing and easy to use. After the design is complete, developers build the code and incorporate the necessary features before carrying out thorough testing to get rid of errors and improve performance.

Mobile applications are mostly distributed through app stores, the most well-known of which are the Google Play Store and Apple's App Store. These platforms act as gatekeepers, limiting app visibility and availability through strict ranking algorithms and approval procedures. The app review process stops harmful software from spreading by ensuring that submitted programs adhere to security and quality criteria. Other distribution methods exist, though, like direct downloads from developers' websites and third-party app stores, which can be especially useful in areas where official shops are not easily accessible.

A vital component of the mobile app ecosystem is monetization, as developers look for long-term income streams to fund further innovation and development. Paid apps, freemium business models, in-app purchases, subscriptions, and ads are just a few monetization techniques. Due to its capacity to draw in a sizable user base before turning them into paying customers, the freemium business model—in which basic capabilities are provided without charge while premium features demand payment—has grown in popularity. Users can purchase virtual products through in-app purchases, especially in gaming applications, while apps that supply ongoing content or services can generate recurrent revenue streams through subscription-based models. With the help of advertising systems like Facebook Audience Network and Google AdMob, developers may make money by showing users advertisements.

A key factor in an app's long-term success is user engagement and retention. In addition to drawing users in, an application should promote recurring use. Personalized experiences, push alerts, gamification, and social media integration are some methods for raising user engagement. Artificial intelligence and data analytics are used in personalization to customize recommendations and content for each user. Although too many push notifications may cause users to uninstall the program, they act as reminders to get them back to it. Gamification adds features like leaderboards, challenges, and awards to make the software more dynamic and interesting. By integrating social media, users may invite friends and discuss their experiences, which promotes natural growth through word-of-mouth advertising.

Given the enormous volumes of financial and personal data saved in mobile applications, security and privacy issues are critical to the ecosystem of mobile apps. Both consumers and developers are in danger from cybersecurity threats like malware infections, phishing scams, and data breaches. Strong security measures, such as end-to-end encryption, safe authentication techniques, and adherence to data protection laws like the California Consumer Privacy Act (CCPA) and the General Data Protection Regulation (GDPR), must be implemented by app developers. However, users need to be careful about the rights they give apps because giving them too much access can result in data exploitation.

Future developments in technology and new trends also impact the mobile app ecosystem. Mobile apps increasingly incorporate machine learning and artificial intelligence to improve features like voice assistants, image recognition, and predictive analytics. Because they offer immersive experiences, augmented reality (AR) and virtual reality (VR) are becoming increasingly popular in gaming, retail, and healthcare applications. The emergence of 5G technology promises reduced latency and higher internet speeds, opening the door for creating increasingly complex and data-intensive apps. Furthermore, the increasing use of blockchain technology in mobile applications improves decentralized transactions, security, and transparency.

The ecosystem of mobile apps is still impacted by laws and regulations enforced by governments and app store operators. App marketplaces regularly revise their rules to address concerns about competitiveness, app quality, and user privacy. Stricter data protection regulations are being introduced by governments worldwide, compelling app makers to prioritize privacy. The continuous legal disputes over app store commissions and antitrust issues highlight the power dynamics between digital companies and developers, which impact the ecosystem's economic environment.

The mobile app market is extremely competitive, with millions of apps competing for users' attention. App store optimization, or ASO, is one of the most important tactics for raising an app's exposure and ranking in search results. ASO optimizes user reviews, app descriptions, keywords, and visual assets to attract new users. Additionally, applications can reach their target audience more successfully using performance marketing strategies like influencer partnerships, social media campaigns, and sponsored ads.

Analytics for mobile apps are essential for comprehending user behavior and enhancing app functionality. To monitor important data like user acquisition, retention rates, session duration, and revenue generation, developers employ analytics solutions like Google Firebase, Mixpanel, and App Annie. Developers can pinpoint problems, improve user experience, and hone monetization tactics through data-driven decision-making. Specifically, A/B testing allows developers to test various app features and layouts to see which consumers respond most.

Since cloud computing offers a scalable infrastructure for hosting apps and managing data, its contribution to the mobile app ecosystem cannot be understated. Developers can delegate processing duties to cloud-based solutions, allowing apps to run smoothly on various devices. Strong backend support is provided by services like Microsoft Azure, Google Cloud, and Amazon Web Services (AWS), allowing apps to manage massive datasets and high traffic levels without sacrificing performance.

Ethical issues about app creation and use are becoming increasingly significant as the mobile app ecosystem develops. The accountability of developers and platform operators is called into question by issues with digital addiction, disinformation, and the moral application of artificial intelligence. A healthier app ecosystem results from initiatives to support ethical app design, such as content filtering tools and features that support digital well-being.

The mobile app ecosystem's worldwide reach demonstrates its influence on various sectors, including healthcare, education, e-commerce, and finance. Mobile applications have revolutionized traditional business models, allowing businesses to access clients across geographic boundaries and provide digital services. While e-learning platforms have made education more accessible, the emergence of fintech apps has transformed banking and payments. Because telehealth apps offer remote monitoring and virtual consultations, they have improved healthcare delivery, especially in rural areas.

There are still obstacles that developers and stakeholders must overcome despite the mobile app ecosystem's enormous growth and potential. New apps find it challenging to acquire momentum due to market saturation, necessitating creative user acquisition and differentiation strategies. Technical problems such as device fragmentation, compatibility problems, and frequent OS changes make app creation and maintenance more difficult. Furthermore, users' expectations always rise, calling for seamless, high-performing apps.

The mobile app ecosystem is a complex digital landscape influenced by user behavior, market dynamics, and technological developments. Every element of the ecosystem, from app development and distribution to monetization and security, is vital to the success of an app. The mobile app market continues to be both difficult and exciting due to the ongoing development of mobile technology, shifting laws, and shifting customer preferences. As mobile applications continue to revolutionize how people engage with the digital world, developers, companies, and users must thoroughly understand this ecosystem.

## Development Approaches and Tools

Development techniques and tools offer frameworks and methodologies for planning, creating, testing, and managing software applications, which are crucial parts of the software engineering process. Rapid technological advancement and the growing complexity of software projects require developers to use various tools and adopt methodical approaches to increase productivity, guarantee high-quality results, and satisfy customer needs. Different methods and resources have been developed over time to meet particular demands in the development process. To shed light on the various techniques and tools utilized in the software development lifecycle (SDLC), including Waterfall, Agile, DevOps, and others, as well as the tools that enable these methodologies, this section will examine these approaches and tools in detail.

Traditional approaches, like the Waterfall model, were common in the early days of software development. The Waterfall model, which is sequential and linear, is frequently regarded as the original formal approach to software development. Gathering requirements, system design, development, testing, deployment, and maintenance are just a few steps in this method's well-defined and organized flow. Without going back to earlier stages, the project proceeds to the next step after finishing each one. Although Waterfall's planning simplicity and clarity are advantageous, its adaptability is limited. If needs change or unanticipated problems emerge later on, the inability to go back and make changes to earlier phases may cause issues. The Waterfall model's shortcomings became apparent as software development advanced, creating more adaptable and iterative approaches like Agile.

Agile development emphasizes adaptability, teamwork, and software iteration and is a substantial departure from conventional methods. Agile promotes iterative cycles called sprints and ongoing feedback, enabling more frequent releases and upgrades. One of its fundamental tenets is Agile's emphasis on adapting to change rather than sticking to a set plan. Because of this flexibility, teams can adjust criteria and enhance the final product in response to user input and changing market demands. Agile ensures that the development process is transparent and responsive to all parties' needs by encouraging tight collaboration between developers, stakeholders, and end users. Working software, customer collaboration, and individual contacts are prioritized over inflexible procedures and tools in the Agile Manifesto, which describes the ideals and principles of Agile. Crucial Agile frameworks, like Extreme Programming (XP), Scrum, and Kanban, provide certain roles and practices to help Agile teams produce high-caliber software.

For instance, one of the most popular Agile frameworks is Scrum. It breaks the development process into several sprints, fixed-length iterations that usually span two to four weeks. Delivering a potentially shippable software increment is the main goal of each sprint. Scrum strongly emphasizes important roles with specific duties, like the Development Team, Product Owner, and Scrum Master. As a facilitator, the Scrum Master ensures the team adheres to Scrum procedures and eliminates any roadblocks that could prevent them from moving forward. The Product Owner prioritizes features and requirements in the product backlog on behalf of the customer. Delivering the work inside the sprint is the responsibility of the development team. Scrum also incorporates routine practices that keep the team together and advancing over time, like sprint planning, daily standups, sprint reviews, and retrospectives.

Another Agile paradigm, Kanban, emphasizes restricting work-in-progress (WIP) and visualizing the process to increase productivity and eliminate bottlenecks. Kanban is more adaptable than Scrum, which operates in sprints with set durations and may be applied in continuous, ongoing development settings. Kanban uses cards on a board to indicate activities that advance through several stages of development, from "to do" to "in progress" to "done." Teams may guarantee more seamless transitions between phases, spot possible problems early, and streamline their workflow by controlling work-in-progress and visualizing the task flow.

Another Agile style that aims to improve software quality and responsiveness to evolving requirements is Extreme Programming (XP). XP strongly emphasizes techniques like frequent releases, pair programming, test-driven development (TDD), and continuous integration. One of XP's main tenets is that developers and clients should collaborate closely to ensure the program satisfies their needs. Because XP promotes design simplicity, developers can concentrate on creating the most important features and avoid superfluous complexity. Writing tests before writing code is known as test-driven development, which helps guarantee that code is dependable and functional. Pair programming encourages knowledge exchange and the early detection of possible problems in the development process by having two developers collaborate on the same computer.

Although many software development teams now use Agile processes, alternative approaches, such as DevOps, have gained popularity, particularly regarding operations and deployment. To close the gap between development and operations teams, DevOps is a collection of methods and a cultural shift that encourages cooperation, automation, and continuous delivery. DevOps highlights how crucial it is to combine development and operations to speed up software delivery, lower mistake rates, and improve product quality. DevOps promotes techniques that allow for quicker and more dependable releases, such as infrastructure as code (IaC), continuous integration (CI), and continuous deployment (CD).

The process of automatically merging code changes into a common repository several times daily is known as continuous integration. By doing this, teams can ensure the product continues to work as new features are added and spot integration problems early. Continuous deployment expands on constant integration and improves deployment speed and reliability by automatically releasing new code to production after it has passed automated tests. Managing infrastructure using code, or "infrastructure as code," facilitates consistent and repeatable system provisioning, configuration, and scaling. In DevOps environments, tools like Jenkins, Docker, Kubernetes, and Terraform are frequently used to automate different parts of the software development and deployment lifecycle.

Software testing is another crucial development component to guarantee the end product's dependability and quality. Different testing methodologies are used to validate various software components and elements, such as user acceptance testing (UAT), system testing, integration testing, and unit testing. Verifying specific code units, like functions or methods, to ensure they operate as intended is the main goal of unit testing. Integration testing verifies that several systems or components interact with one another in a way that makes sense. System testing entails evaluating the application to ensure the program satisfies the requirements. Before the program is put into production, user acceptance testing—usually done by end users—ensures it fulfills their requirements and expectations.

Numerous solutions are available to automate and streamline the testing process. One popular technology for automating web browser testing of web applications is Selenium. Two well-liked frameworks for unit testing Java programs are JUnit and TestNG. Postman and SoapUI are two popular tools for testing APIs in integration testing. Automated testing technologies are included in continuous integration and continuous delivery (CI/CD) pipelines to guarantee that tests are executed automatically anytime the codebase is modified.

Another crucial component of contemporary software development is version management. It enables developers to work together, monitor codebase changes, and roll back to earlier iterations as needed. GitHub and

GitLab offer cloud-based repositories for code management and collaboration, making Git one of the most widely used version control systems. Developers can build branches with Git to work on bug fixes or new features without affecting the main source. Pull requests allow other team members to examine the code before it is integrated, allowing a feature or patch to be merged back into the main branch once finished.

Another essential tool in the software development process is an integrated development environment or IDE. IDEs give programmers capabilities and tools to write, test, and debug code more effectively. Features like code completion, syntax highlighting, debugging tools, and version control integration are available in well-known IDEs like Visual Studio, IntelliJ IDEA, and Eclipse. Developers may write code more quickly, find mistakes early, and streamline the process using these tools.

Lastly, project management tools are essential for planning and monitoring the advancement of development projects. Software development projects frequently employ Jira, Trello, and Asana as tools for task management, issue tracking, and progress tracking. These tools facilitate efficient communication, task delegation, deadline setting, and team organization. Project management tools are frequently used in Agile contexts to plan sprints, manage product backlogs, and monitor user story fulfillment.

To sum up, the software development process relies heavily on development methodologies and technologies. Every methodology, from contemporary approaches like Agile and DevOps to more conventional models like Waterfall, has special benefits and tackles various issues. Development teams may increase cooperation, boost productivity, and guarantee the delivery of high-quality software by using the proper tools and developing the right methodology. The landscape of software development methodologies and tools will surely continue to shift as technology advances, with new techniques and inventions appearing to satisfy the needs of a constantly changing sector.

## Setting Up the Development Environment

One of software development's most important first steps is setting up a development environment. It can greatly impact output and the development process's performance. The procedure entails setting up the software, tools, and technologies required for developers to write, test, and release their code efficiently code efficiently. Regardless of the software project size—from a tiny personal application to a major enterprise solution—it is an essential first step. Although setting up a proper development environment may initially appear like a difficult effort, it may facilitate and organize the development process once configured. The significance of a development environment, how to set it up, the required tools, and the factors developers should consider will all be covered in this section.

Knowing the software project's needs is the first thing to consider while establishing a development environment. Depending on the programming languages, frameworks, and technologies being used, various projects call for different configurations. For instance, a web application project may require a browser for testing, Node.js, and a code editor such as Visual Studio Code. In contrast, a mobile application would need Xcode or Android Studio. Since it dictates which software and utilities must be installed on the system, it is crucial to comprehend these requirements. It's simple to install extraneous tools when project requirements are unclear, which can cause confusion, time wastage, and possible compatibility problems.

Selecting the appropriate hardware is the next stage. The minimal specifications needed to operate the required tools and software should be met by the machine or server on which the development environment is being set up. For example, certain operating systems or more potent hardware may be necessary for the best performance of particular programming languages or frameworks. For instance, a computer with enough RAM and processing power is essential while working with Java, particularly when performing resource-intensive operations like compiling sizable codebases. Additionally, it's critical to use an operating system that supports the development tools. Operating systems based on Linux, macOS, or Windows are available to developers. Although every operating system has pros and cons, it's crucial to make sure it works with the software stack being utilized.

The software and tools that will make up the development environment's core must be chosen after the hardware and operating system have been decided. The code editor, often known as the Integrated Development Environment (IDE), is the most crucial tool for any developer. To assist developers in writing and debugging code, IDEs are specialized tools. Eclipse, JetBrains IntelliJ IDEA, and Visual Studio Code are well-known IDEs. The coding process can be significantly accelerated by the capabilities these tools provide, which include syntax highlighting, auto-completion, error checking, and integration with version control systems. Furthermore, they frequently offer plugins or extensions that let developers incorporate extra languages, frameworks, and tools according to their requirements.

Other tools must be installed in addition to the IDE, depending on the project's requirements. For instance, team collaboration, code change management, and maintaining a clean change history are all made possible by version control systems like Git. Git allows developers to keep track of code changes, roll back to earlier iterations, and work together more efficiently. Services that host repositories and offer extra collaboration capabilities, such as GitHub, GitLab, or Bitbucket, are essential for contemporary software development. To work with repositories efficiently, developers should install a Git GUI client or the Git command-line tool.

Configuring the programming languages and frameworks needed for the project is another crucial step in creating a working environment. Each language, such as Python, JavaScript, Java, or Ruby, requires package managers, environments, and setup processes. For instance, while working with Python, developers must install both Python and a package manager such as pip, which makes it simple to install dependencies and third-party libraries. Similarly, tools like Node.js, npm, or Yarn may be required while working with JavaScript. Certain configurations are also necessary for frameworks like Django, React, or Angular, such as setting up environment variables and installing dependencies. By offering pre-built components and ready-to-use functionality, these tools and frameworks assist developers in streamlining their workflow and creating apps more quickly.

Configuring the development environment also includes setting up databases. A database is essential to the environment since many software programs need to store and access data. Depending on the project's needs, developers frequently utilize relational databases like MySQL or PostgreSQL or NoSQL databases like MongoDB. Installing and configuring the required database software is the responsibility of developers. For easier database interaction, developers should also have a database management tool, such as pgAdmin for PostgreSQL or phpMyAdmin for MySQL. Configuring database access, including establishing users, permissions, and secure credentials, is also crucial.

After installing the necessary software and tools, it's time to configure the rest of the development environment's components. For instance, developers frequently need to build virtual machines or containers to replicate production environments. Using virtualization software such as Docker or VMware, developers can establish isolated environments to test their code without affecting the main system. With the help of these tools, developers can duplicate the production environment and ensure the program operates as intended in various settings. Docker, for example, is well-known for building lightweight containers that bundle the application and its dependencies to ensure consistency across many systems.

A task runner or build automation tool is another helpful tool for configuring a development environment. Common development operations like code minification, test execution, and application deployment are automated by these technologies. Well-known tools like Webpack, Gulp, and Grunt help streamline repetitive processes and increase development workflow efficiency. Automating these processes makes errors less likely to occur, and consistency is guaranteed across development stages.

Developers should consider creating testing environments when installing and setting up development tools. A crucial step in the software development lifecycle, testing ensures that the code functions as intended and errors are found early. To see any problems, testing environments should be as similar to the production environment as feasible. Depending on the programming language, developers frequently utilize frameworks for unit testing such as JUnit, Mocha, or Pytest. To automate the testing process, developers must also ensure continuous integration (CI) pipelines are in place. When developers make changes to the codebase, CI technologies like Jenkins, CircleCI, or Travis CI assist them in running automated tests to make sure the program stays dependable and stable.

Security is a crucial factor to consider while establishing a development environment. Developers are responsible for maintaining a secure environment and protecting sensitive data, including passwords, certificates, and API keys. Sensitive information should be stored using encryption, safe vaults, and environment variables. Sensitive information should not be hardcoded into the source code; developers should manage and retrieve it securely. The development environment can also be shielded from potential vulnerabilities by installing firewalls, restricting network access, and maintaining software updates.

After the development environment is completely configured, document the settings for future use. Other developers on the team may find this documentation useful in understanding how to configure the environment on their computers and guarantee uniformity among various platforms. Installing and configuring all required tools, libraries, frameworks, and databases should be included in the documentation. It should also outline any particular environment variables, access credentials, or configurations needed for the project. Developers can save time and lower the possibility of misunderstandings or mistakes when new team members join the project by having this documentation in place.

Lastly, the upkeep of the development environment must be taken into account. Software versions evolve with time, new tools are created, and outdated tools may be phased out. It is crucial to update the development environment frequently to guarantee compliance with the most recent updates and software versions. Developers should routinely check their environment and update any out-of-date elements, including libraries, frameworks, or programming language versions. By doing this, the environment is kept safe, effective, and able to accommodate the newest development techniques.

A software project's success can be significantly impacted by the crucial work of setting up a development environment. A properly set environment guarantees that the code works reliably across machines, increases development productivity, and lowers the possibility of errors. Choosing the appropriate hardware and software, configuring programming languages and frameworks, creating databases, and ensuring that testing, security, and maintenance protocols are followed are all steps in the environment setup process. Through careful configuration and ongoing maintenance, developers may build a solid basis for their projects and steer clear of pointless problems that could impede progress.

## Programming Languages for Mobile Development

Programming languages are crucial for mobile development because they specify how software is made, works, and communicates with users and other mobile device components. Numerous programming languages have emerged in mobile development, each with unique advantages, disadvantages, and applicability for various applications. The need for specialized, effective, and potent languages to develop a wide range of mobile applications is growing as the mobile industry expands. Nowadays, programming languages used in mobile development range from more recent cross-platform languages that allow developers to construct applications for numerous mobile devices using a single codebase to more traditional, platform-specific languages. Developers who want to produce mobile applications that are effective, user-friendly, and of high quality must have a solid understanding of these languages and their uses.

Platform-specific programming languages were the main emphasis of mobile development in the past. Platform-specific languages, like Java for Android and Objective-C for iOS, dominated the early days of mobile app development. These programming languages were created to fully utilize their corresponding operating systems' special qualities and functionalities, giving programmers the means to produce native, high-performance programs. Cross-platform development became increasingly necessary as the mobile market changed, and new programming languages and frameworks were created to enable developers to write code once and have it run on several devices.

Java is one of the most popular and extensively utilized programming languages for creating mobile applications. Since the beginning, it has been the main language used to develop Android apps. Java has long been at the center of the developer community for Android, the world's most widely used mobile operating system. Because Java is an object-oriented language, developers can design modular, maintainable, and scalable apps, making it perfect for mobile development. Furthermore, Java offers a wealth of libraries and APIs that let programmers include cutting-edge functionalities like networking, multimedia, and user interface elements into their apps.

The development of Android also heavily relies on the Java Virtual Machine (JVM). Bytecode, which can execute on any device with a JVM, regardless of the underlying hardware, is created by compiling Java code using the JVM. This capability gives Java great portability and cross-platform compatibility, which is very beneficial for mobile developers. Java is a fundamental language for mobile development due to its long-standing dominance in the Android ecosystem. However, newer languages like Kotlin are gradually replacing it.

Kotlin, a more recent programming language, has become very popular in the Android development community. Since Google declared Kotlin an officially supported language for Android development in 2017, many Android developers have swiftly adopted it as their preferred language. Because Kotlin and Java are completely compatible, developers can use Kotlin in addition to pre-existing Java codebases. One of Kotlin's primary benefits is its expressive and succinct syntax, which enables programmers to accomplish the same task with less code. This is a more efficient option for many developers because it can result in shorter development cycles and fewer defects.

Additionally, Kotlin adds several features that enhance Java, like null safety, which helps avoid null pointer exceptions, one of the most frequent causes of defects in Java applications. Kotlin is a more contemporary and adaptable option compared to Java due to its support for functional programming elements like immutability and higher-order functions. These characteristics and Kotlin's ability to interact with pre-existing Java libraries and frameworks have made it a desirable choice for developing Android apps. Consequently, many Android developers are switching from Java to Kotlin, and it is anticipated that Kotlin will continue to gain traction.

Objective-C served as the main programming language for iOS mobile development for many years. The preferred language for Apple's macOS and iOS systems is Objective-C, an object-oriented language created in the early 1980s. C code is likewise valid in Objective-C since it is a superset of the C programming language. The messaging syntax of Objective-C, which enables objects to communicate with one another, is one of its primary characteristics. This capability aids developers in producing extremely adaptable and modular applications, which is crucial for mobile development.

However, learning Objective-C can be challenging, particularly for developers unfamiliar with general programming or mobile development. It can have complicated syntax and necessitate a thorough knowledge of memory management, which, if not managed correctly, can result in problems like memory leaks. Thus, Swift, a more recent and approachable programming language released by Apple in 2014, has progressively supplanted Objective-C.

Swift was created to fix many Objective-C issues while still being compatible with current Objective-C codebases. Swift is a more user-friendly, type-safe, and memory-safe language than Objective-C. It is especially well-suited for mobile programming since it has numerous performance-enhancing features, like automated memory management and multi-threading optimizations. Swift's clear and simple syntax facilitates code writing and maintenance for developers, and its seamless integration with the iOS and macOS frameworks offers a smooth development experience for Apple platforms.

Strong features like closures, which enable functional programming techniques, and options, which aid in handling null or missing data more securely, are also included in Swift. Swift is the go-to language for developing iOS and macOS apps because of its emphasis on security and performance. Swift is currently regarded as the future of mobile development for Apple devices, having quickly supplanted Objective-C as the preferred language for Apple developers.

Cross-platform development frameworks and languages are becoming increasingly popular, even if platform-specific languages like Java, Kotlin, Objective-C, and Swift have dominated the mobile development scene. Without requiring distinct development efforts for each platform, cross-platform development enables developers to create a single codebase that can be delivered on several platforms, including iOS and Android. This strategy is more economical and effective for companies that wish to reach a large audience while preserving a consistent user experience across devices.

Along with frameworks like React Native and Ionic, JavaScript is one of the most widely used cross-platform languages. The most common language for web development is JavaScript, which has also gained popularity for mobile development thanks to frameworks like React Native. With React Native, developers can utilize JavaScript and React, a well-known user interface toolkit, to create mobile applications. React Native preserves the advantages of a single codebase by compiling JavaScript code into native components, enabling developers to construct mobile applications with a native appearance and feel.

React Native's near-native performance is one of its main advantages; as a result, applications developed with React Native are frequently identical to those created in platform-specific languages like Swift or Kotlin. React Native is, therefore, a strong option for cross-platform smartphone programming. Furthermore, there is a sizable and vibrant React Native developer community, making it simple for developers to locate tools, libraries, and plugins to increase the functionality of their apps.

Flutter is another well-known cross-platform framework that Google created. The Dart programming language, designed especially for building scalable, high-performance apps, can be used to develop applications with Flutter. Instead of depending on native components, Flutter employs a novel rendering technique in which the user interface is drawn directly into the screen. Because of this, Flutter apps can be highly customized and are consistent across platforms. Flutter has become a hugely popular framework in recent years because of its quick development cycle, expressive user interface, and compatibility with iOS, Android, and the web.

Other cross-platform frameworks include Apache Cordova (previously PhoneGap), which enables developers to create mobile applications using HTML, CSS, and JavaScript, and Xamarin, which uses C# and the.NET framework. Despite being less popular than React Native and Flutter, these frameworks are useful for certain use cases and provide several benefits for cross-platform development.

To sum up, many different programming languages are available in mobile development, each serving a distinct set of requirements and tastes. While cross-platform frameworks like React Native, Flutter, and Xamarin enable developers to write code once and have it run on various platforms, Java, Kotlin, Objective-C, and Swift continue to be the most popular languages for native mobile development. New languages and frameworks will probably appear as mobile development develops, giving developers even more choices when making effective, potent, and intuitive mobile apps. Although the target platform, the application's complexity, and the developer's skill level ultimately determine the programming language to use, there is no lack of tools for creating the upcoming generation of mobile applications thanks to the wide range of possibilities accessible.

# CHAPTER II

# User Interface (UI) and User Experience (UX) Design

## Mobile UI/UX Design Principles

A mobile application's overall success largely depends on its UI/UX design. Users' expectations for mobile app design have changed as mobile devices have grown more commonplace. These days, users expect smooth, simple, and eye-catching experiences. In this regard, developing applications that live up to these expectations requires a solid understanding of mobile UI/UX design principles. The visual components of an application, such as buttons, icons, and layout, are referred to as the user interface, or UI. Contrarily, UX (User Experience) refers to the total app experience, encompassing usability, accessibility, and user happiness. Both of these elements are combined in effective mobile UI/UX design to produce useful, entertaining, and engaging apps for users.

Simplicity is a fundamental design principle for mobile UI/UX. Mobile devices have a small screen area; therefore, ensuring the design is clear and uncluttered is important. Users may become frustrated and stop using the program due to an overly difficult design. In mobile app development, where the emphasis is on the key components that propel the app's operation, a minimalist design approach frequently yields the greatest results. Each component must have a distinct function, and superfluous elements must be avoided. This simplicity applies to the app's navigation, structure, and visual appeal. Users may easily and swiftly accomplish their goals with a simplified user flow and clear, understandable navigation.

Another essential tenet of mobile UI/UX design is consistency. A consistent design framework guarantees a unified and user-friendly app. This uniformity extends to interactive aspects and visual components like button designs, colors, and fonts. Users should have a consistent experience interacting with the app's various screens or functionalities. Buttons, for instance, ought to appear and function uniformly on different panels. The user can anticipate how the app will react and what to expect while interacting with it because this consistency helps them create a mental model. Consistency also applies to the brand's identification, ensuring the app's design complements the company's colors, emblems, and voice.

Responsiveness is another important UI/UX design paradigm for mobile devices. The app's design must adjust to various screen sizes and orientations because many different mobile devices are in use today, from tiny smartphones to bigger tablets. Thanks to its responsive design, the software will continue to be usable and aesthetically pleasing on any device. To ensure a consistent experience across several devices, this adaptability entails changing the layout, resizing graphics, and moving items as needed. Apart from screen size, responsiveness also describes how the application responds to various interaction methods, such as voice instructions or touch gestures. An application that reacts to user inputs quickly and fluidly improves the user experience overall and helps users avoid dissatisfaction.

One of the most crucial tenets of mobile UI/UX design is user-centered design. This principle highlights the importance of comprehending the intended user base and creating the app with their requirements and preferences in mind. To produce a user-centered design, designers must conduct in-depth research to understand their users' objectives, expectations, and pain areas. Usability testing, interviews, and user surveys may be used in this study. To guarantee that the app addresses actual issues for its users, the design process should be guided by the knowledge gathered from this study. Beyond functionality, user-centered design makes sure that the app offers a pleasurable experience that emotionally connects with users and encourages them to return.

In mobile UI/UX design, the accessibility principle is equally essential. As the user base grows more varied, developers need to ensure that users of different skill levels can use their programs. High-contrast color schemes, text-to-speech capabilities, font size adjustments, and other assistive technologies are accessibility features that help people with impairments navigate an application. By implementing these features, designers make sure their apps are inclusive and usable by a wider range of users. In addition to its social function, accessibility can increase the app's user base and attractiveness to a wider range of users.

Two key ideas that can improve the user experience are interaction and feedback. Whether swiping, tapping a button, or completing a form, mobile users anticipate instantaneous reactions to their activities. Giving consumers prompt, unambiguous responses lets them know that the app takes their input seriously. Visual signals like button animations, loading indicators, or confirmation messages can do this. Additionally, feedback should be instructive, assisting users in comprehending the outcomes of their choices. When a user submits a form, for instance, the application ought to offer feedback to let the user know if the submission was successful or if there were any issues that needed to be fixed. There should be no discernible lag or response time delays, and interactions should be fluid and seamless. Users' experiences may suffer if the app doesn't give them this feedback, leaving them perplexed or irritated.

Navigation is one of the most important components of mobile UI/UX design. There must be no difficulty for users to navigate the app. Users better understand where they are in the app and how to reach their objective with clear and simple navigation. To facilitate user exploration, mobile applications frequently employ a variety of navigation patterns, including side menus, bottom navigation bars, and swipe movements. The navigation system's design should consider the most crucial elements and activities to ensure they are accessible. The main objectives that users wish to accomplish should also be reflected in the logical order of navigation components. Users may become frustrated and stop using the app due to complicated or concealed navigation.

Another important component of mobile UI/UX design is performance optimization. Mobile apps must function well since unresponsive or sluggish apps can alienate consumers. Careful coding, effective resource management, and reducing pointless operations that deplete the device's memory or battery can all lead to optimization. There should be little latency between interactions or loading times in the application. In addition to enhancing the user experience, performance optimization makes the app stand out in a crowded market where consumers have many options and little tolerance for sluggish or subpar apps.

Using a touch-friendly design is another crucial idea. In contrast to desktop apps, mobile apps mostly rely on touch gestures like pinches, swipes, and touches. Touch-friendly interface design guarantees that users may engage with the application comfortably and naturally. For ease of tapping, buttons, links, and other interactive components should be sized correctly. Errors and dissatisfaction may result from small or closely spaced components. The ergonomics of using a mobile device are also considered by touch-friendly design, which guarantees that the app is comfortable to use whether the user is holding the device in one or two hands. Important interactive components, like buttons or menus, should be positioned with ease of access and with consideration for typical user behaviors.

Another important aspect of mobile UI/UX design is aesthetics and images. An app's visual attractiveness can make or break it since users are more inclined to interact with aesthetically pleasing and well-designed apps. But utility shouldn't be sacrificed for beauty. The design should emphasize legibility and clarity using suitable fonts, colors, and contrast. Icons and pictures are visual components that should be user-friendly and consistent with the app's general concept. A unified visual style fosters professionalism and trust, which motivates users to interact with the app more fully. Along with supporting the app's functionality, the visual design should let users engage with it in a natural and intuitive way.

In today's mobile-centric world, it is also crucial to incorporate mobile-first design ideas. Instead of reducing a desktop design to fit mobile screens, the mobile-first design creates an app or website with mobile consumers in mind first. Because more people use mobile devices to browse the internet and utilize apps, the mobile-first design ensures that the app works well on smaller screens. Additionally, designers are forced to prioritize the most important features and content because there is less space for unnecessary items. Developers can produce optimized experiences for mobile use by concentrating on mobile-first design, eventually improving the user experience.

Lastly, the mobile UI/UX design process is incomplete without testing and iteration. An app must be extensively tested after development to ensure it works as planned and satisfies user needs. This testing might involve performance testing to ensure the app functions properly on various devices and network settings and usability testing, in which actual users engage with the app and offer input on its design. The design should be improved and refined based on the feedback received during testing. Before making the app available to a wider audience, designers can find and address issues through iterative testing and improvement.

To sum up, mobile UI/UX design is a complicated and multidimensional process that requires careful balancing between user experience, functionality, and visual design. Through adherence to fundamental concepts like accessibility, responsiveness, consistency, and simplicity, designers may produce mobile applications that are not only efficient but also pleasurable to use. Mobile apps that prioritize user-centered design, intuitive navigation, performance optimization, and touch-friendly interfaces are guaranteed to satisfy user demands and differentiate themselves in a market that is becoming increasingly competitive. Creating applications that consumers like using is the ultimate goal of mobile UI/UX design, as this promotes sustained engagement and satisfaction.

## Creating and Structuring App Interfaces

A key component of app development is designing and organizing the user interface, which calls for carefully considering both design principles and user experience (UX). An application's interface is where users interact with the program. This interface's structure and design are essential to ensuring the application is useful, easy to use, visually appealing, and accessible. The design of interfaces has changed over time, and the move toward user-centric design has made it more crucial than ever to take end-user wants and behaviors into account. From initial concept development to the final product launch, there are several steps involved in creating and organizing app interfaces, and it requires knowledge of visual design, interaction design, and usability standards.

Knowing the app's goal and target user base is the first step in designing a successful user interface. Identifying the app's main users and defining its objectives is crucial for directing the design process. For example, an e-commerce or social media platform will have a different interface than a fitness monitoring app. Usability can be enhanced by creating an interface that meets the specific requirements of each kind of program. Design choices will also be further informed by knowledge of user demographics, including age, tech proficiency, and preferences. For instance, an older audience might value readability and simplicity, whereas a younger one would value vibrant, engaging designs.

The next stage is concentrating on the user experience (UX) design after determining the app's goal and target user base. Ensuring the app is user-friendly, effective, and enjoyable to use is the focus of UX design. Its main goal is to comprehend user behaviors and needs to design an interface that meets those objectives. Developing user personas—detailed depictions of the many user types who might engage with the application—is one of the essential elements of UX design. It is simpler to create an interface that is both user-friendly and satisfies the demands of target users when designers are able to visualize their goals, challenges and wants thanks to these personas.

Another crucial tool for organizing application interfaces is the user flow diagram. These diagrams show the user's journey through the application to complete a particular job. User flow diagrams assist designers in making sure that the app's navigation is rational, easy to use, and effective by describing the different processes and interactions involved in an activity. Thanks to a well-designed user flow, users thanks to well-designed users, users may rapidly learn how to use the app and obtain the required features. This guarantees that the app will remain functional even as the number of features grows, which is especially crucial for apps with numerous features or intricate procedures.

During the subsequent stage of interface development, wireframing becomes crucial. Basic, low-fidelity drawings or blueprints of the application's user interface are called wireframes. They are used to build out the basic framework of each screen and usually lack intricate graphics. With wireframing, designers can see how the application will be organized, including placing various components like buttons, menus, and text. The app's hierarchy and structure are roughly depicted, enabling the designer to experiment with different combinations and layouts before moving forward with more intricate design work. Because they allow stakeholders to comprehend the app's structure rapidly and offer suggestions for enhancements, wireframes are particularly helpful for obtaining input early in the design process.

Following the completion of wireframes, attention is turned to visual design, which entails incorporating images, typography, colors, and other graphic components to produce a visually appealing interface. The user's initial impression of the application is greatly influenced by visual design. Thus, ensuring the interface is aesthetically pleasing and practical is critical. For instance, colors can convey vital information, such as status (red for errors or alarms), and elicit particular feelings or moods. Another important visual design component is typography, which influences readability and can contribute to the app's overall tone. The app's UI is neat, polished, and user-friendly by consistently using fonts, spacing, and alignment.

Selecting the appropriate interaction design components is another step in developing an application interface. The main focus of interaction design is how users will interact with the app—including touch motions, button locations, and screen transitions. Reducing the cognitive burden needed for users to navigate the app is one of the objectives of interaction design. Important activities, for instance, should be quickly accessible with only a few taps or clicks, and buttons should be positioned in prominent areas. Another aspect of interaction design is creating visual feedback to let users know when an activity is successful or unsuccessful. For example, a progress bar or loading spinner can help consumers gauge how long a process will take to finish.

It is crucial to consider mobile-specific factors like screen size and touch-based interactions as the app design develops. Because mobile devices' screens are smaller than desktops or laptops, app interfaces must be made with this restriction in mind. Designers must prioritize the most important features and content to keep the app usable and navigable on smaller screens. This frequently entails ensuring the interface is clutter-free and applying minimalist design concepts. Larger touch targets for buttons and links are also necessary for touch-based interactions since users need to be able to tap items precisely without becoming frustrated. Additionally, the interface should support popular mobile device movements like pinching, tapping, and swiping.

Ensuring the design is the same throughout the program is another crucial component of the app interface structure. Users can anticipate how various elements will behave and become more accustomed to the app when the design is consistent. Using the same color palette, font, and layout patterns on all displays will help achieve this. To avoid having to relearn how to use the app on each screen, buttons and navigation menus, for instance, should be placed consistently throughout the application. Consistency in design is frequently achieved through the use of design systems, which are detailed standards that specify the visual and interaction patterns for an application. These systems streamline the design and development process, giving designers and developers a single point of reference.

An essential step in the interface design process is user testing. Testing the interface with actual users after it has been designed and organized is crucial to finding any problems or potential areas for enhancement. Usability testing entails seeing users engage with the application and getting their opinions. Before the app is made available to a wider audience, designers can make necessary adjustments by using this testing to find portions of the app's UI that cause confusion, annoyance, or inefficiency. Another useful technique for improving the app's design is A/B testing, which compares two interface iterations to see which works better.

Accessibility is a crucial factor to take into account when designing app interfaces. Ensuring an app is usable by everyone, including people with impairments, is a basic component of excellent design. This entails designing user interfaces that people with hearing, vision, or mobility disabilities can use without difficulty. To enable screen readers to read photos aloud, designers should ensure that text is large enough for individuals with visual impairments to read and that images include alternate text descriptions. Because some color combinations can be hard to discern, color contrast is especially crucial for users who are color blind. Designers may produce apps that are usable by a larger spectrum of users by putting accessibility first, enhancing the overall user experience.

Performance optimization must be considered while designing and structuring the UI as the app is developed. If an app is sluggish or unresponsive, even the most exquisitely designed one will be inadequate. Optimizing image sizes, cutting back on superfluous animations or transitions, and lowering the quantity of HTTP requests the application makes are all aspects of performance. To prevent users from becoming irritated by delays, the interface should be made to load rapidly, even on slower network connections. Performance testing is crucial to find and fix any bottlenecks that can impair the app's responsiveness or speed.

Lastly, interface maintenance is a continuous activity after the app is published. Designers may need to update or enhance the interface in response to changing user needs. Adding new features, changing the design, or improving the user flow in response to feedback could all be part of this. User engagement and app performance metrics can be continuously monitored to reveal areas that could require improvement. Over time, the app will continue to be useful, current, and easy to use, thanks to frequent updates and enhancements.

To sum up, developing and structuring app interfaces is a complex process that includes using the right tools and approaches, knowing the target audience, and designing with usability in mind. The procedure entails establishing the objectives of the application, developing user personas, organizing user flows, wireframing, producing aesthetically pleasing interfaces, and testing the design to ensure it satisfies user requirements. A successful app interface also must have performance optimization, accessibility, consistency in design, and considerations for mobile devices. Developers may produce useful but also entertaining and simple apps by concentrating on these guidelines and iteratively enhancing the interface in response to user input, guaranteeing a satisfying experience for every user.

## Navigation and Interaction Design

Navigation and interaction design are essential to create intuitive, user-friendly interfaces that improve the entire user experience of digital products. Knowing the fundamentals of navigation and interface design is more important than ever in the current online and app design era, where customers expect seamless experiences and ease of use. By establishing a communication channel between the user and the system, both components make it possible for the user to engage with the interface in a productive and pleasurable way. While interface design seeks to make such gestures fun and meaningful, navigation design aims to enable users to navigate the system easily. They serve as the cornerstone of a well-designed interface that can accommodate the usability and functionality that contemporary users need.

Navigation design is how users navigate a digital product, whether a website, mobile application, or software. It includes the arrangement, arrangement, and structure of buttons, menus, links, and other components that direct users between sections. A seamless and intuitive experience is promoted by efficient navigation, which makes it easier for consumers to locate what they're looking for. Logical, consistent, and well-organized navigation is necessary to do this. Information must be arranged clearly and hierarchically, with categories the user can understand. A website that sells goods, for example, might arrange categories like "Electronics," "Clothing," and "Accessories" according to the most typical user behavior. As a result, users can easily and swiftly browse to the section that piques their interest.

The first steps in the navigation design process are finding the system's main objectives and comprehending user needs. Regarding websites, this sometimes entails examining customers' most frequent actions, such as perusing merchandise, making purchases, or reading material. By comprehending these behaviors, designers may prioritize these tasks in their navigation framework. Furthermore, the navigation must be responsive to various screen sizes and devices. Given the growing popularity of mobile devices, navigation needs to function flawlessly on multiple platforms and provide a responsive design that changes to fit the screen's size and orientation. This could entail using mobile-first design techniques that maximize the user experience on smaller devices, such as hidden navigation bars or collapsible menus.

Simplicity is a fundamental principle in navigation design. Consumers may become frustrated and leave the website or app when a navigation system is overly complicated. Designers must, therefore, balance offering a sufficient number of alternatives with maintaining an easy-to-use and clutter-free navigation system. Confusion can result from having too many options on one page, and users may find it challenging to do their jobs if there are too few options. Ensuring that each navigation item has a clear and self-explanatory description is crucial. Making navigation intuitive requires using well-known terminology that fits users' mental models. Clear visual cues like buttons, links, and icons also aid in directing users through the interface.

Conversely, interface design concerns how people engage with the system and how the system reacts to those engagements. It entails creating the functionality of interactive components, including sliders, buttons, forms, and other controls. Interaction design aims to produce intuitive, meaningful experiences that raise user pleasure and engagement. Ensuring the system gives feedback validating user actions is crucial to creating an interactive and responsive experience. Feedback can take many forms, such as haptic feedback on mobile devices, aural feedback, or visual indications like illuminating a button when it is hovered over. These indicators foster system trust by assisting users in understanding the consequences of their activities.

Making sure that interactive features are easy to use and discoverable is one of the most basic parts of interaction design. The interaction gets annoying if the user has trouble finding the necessary buttons or controls or if the controls are hard to understand. Interaction designers employ concepts like signifiers and affordance to address this. The characteristics of an object that imply its purpose—such as a slider that implies sliding or a button that appears clickable—are referred to as affordability. A label next to a form field that informs users of the information needed is an example of a signifier, a visual or verbal signal conveying how an object should be used. Signifiers and affordances help users interact with the system as intended by the designer.

The usage of consistency is a crucial component in interaction design. Consistent design patterns save cognitive burden and enable users to quickly become familiar with the system. This entails using well-known interface components, keeping a consistent design, and ensuring that interactive features work consistently throughout the interface. Buttons should, for example, always look the same and behave the same way—for example, changing color when clicked or hovered over. Similarly, items should be arranged according to well-known patterns, such as positioning navigation menus at the top or side of the screen to save users from seeking them.

Another crucial component of interaction design is responsiveness, in addition to consistency. Today's Consumers demand digital items to respond to their activities instantly, without lag or malfunction. Users may become frustrated and have a worse overall experience if response times are slow. It is crucial to optimize performance to guarantee smooth and fluid interactions. This can be accomplished by using strategies like animation optimization to ensure they function properly on various devices or lazy loading, which postpones the loading of specific parts until they are required.

In interface design, reducing the work users must do to accomplish a job is another crucial idea. The aim is to reduce the cognitive burden and the number of steps required to engage with the system. Users can do activities more quickly and with less mental strain if forms are made simpler by, for example, lowering the number of needed fields, employing sensible defaults, or providing auto-completion. Cognitive friction can also be decreased by creating interactions that mimic human behavior. For instance, adding voice commands to speech-activated devices or gesture-based interactions to mobile apps can improve the user experience by making it feel more natural and responsive to their expectations.

The close relationship between effective navigation and interaction design produces a seamless and pleasurable user experience. While interaction design concentrates on how users interact with the information as they progress through the system, good navigation guarantees that users can easily find their way around it. Users can easily navigate a system, discovering what they need with little effort and engaging with the interface in a comfortable and natural way when these two components are carefully planned and implemented.

Though these guidelines might seem simple, a thorough grasp of user requirements and behavior is necessary to strike the ideal balance between navigation and interface design. Both disciplines heavily rely on user research, with designers using techniques such as surveys, A/B testing, usability testing, and user interviews to learn more about how people engage with a system. Designers can improve and streamline the navigation and interaction patterns by using these insights to pinpoint problems and places that need work.

Usability testing is crucial for evaluating how well navigation and interaction design work. By watching actual users engage with a prototype or live system, designers can spot areas where users struggle, get lost, or fail to finish tasks. The design can then be improved, and this input can fix any problems. For instance, the designer may try experimenting with different positions, sizes, or colors to increase visibility if consumers have trouble finding the checkout button on an e-commerce site. Similarly, designers can change the phrasing or provide more instructions to make the procedure more understandable if users routinely abandon a form because of unclear labels.

Additionally, when creating navigation and interface components, accessibility must be considered. Users with disabilities, such as visual, hearing, or motor impairments, can interact with the system efficiently if it is accessible. The Web Content Accessibility Guidelines (WCAG) and other accessibility guidelines establish a set of best practices for making digital products useable by individuals with various abilities. This entails ensuring that interactive features are keyboard-navigable, offering alternative text for images, and maintaining adequate contrast between text and backdrop colors. Designers may guarantee that their products are inclusive and usable by a larger audience by keeping accessibility in mind.

To sum up, effective and satisfying user experiences depend heavily on navigation and interface design. While interface design ensures that users' interactions are responsive, easy to use, and pleasurable, navigation helps people navigate a system. To satisfy user needs and produce a smooth, pleasurable experience, thorough planning, investigation, and testing are necessary. Designers may create systems that people find simple to use and enjoyable to interact with by emphasizing accessibility, responsiveness, consistency, and clarity. The fundamentals of navigation and interface design will continue to be crucial in developing effective, user-centric products as the digital world changes.

## Theming and Styling Apps

Apps for styling and theming have become a vital component of the mobile experience, allowing users to customize and improve the look of their devices. These applications offer several capabilities that let users alter their smartphones' or tablets' look and feel, from applying unique themes and fonts to altering wallpapers and icons. Individuality and self-expression in digital spaces have become increasingly important as smartphones and other mobile devices have taken center stage in our daily lives. By allowing users to customize their devices to suit their tastes and preferences, theme and style apps make using a device more pleasurable and unique. These apps have become increasingly popular, and iOS and Android devices now provide various customizing possibilities.

Humans have always wanted to customize their surroundings, which is why theming and styling apps have become so popular. Users could personalize their desktop environments in the early days of personal computing by selecting various themes, wallpapers, and icons to give their machines a more unique feel. The same need for personalization permeated the mobile realm as mobile devices proliferated, especially with the introduction of smartphones. But at first, customers' ability to customize their devices was limited by the restrictions placed on them by mobile operating systems. The necessity for theming and styling options increased over time as mobile platforms developed and became more adaptable, which prompted the creation of apps to satisfy this demand expressly.

Android was the first significant mobile operating system to adopt theming and design. As an open-source platform, Android has always been more lenient than its rivals, enabling users to install third-party apps that alter the device's look. With the ability to adjust almost every element of the system's UI, Android's customization features have long been a major selling point for customers. Mobile theming saw a sea change with the advent of launchers, which are applications that let users alter the home screen, app icons, and navigation style. Without having to change the underlying operating system itself, launchers enabled users to design new user interfaces that altered the appearance and feel of their devices. These days, Android users may choose from a wide variety of theming and styling applications, such as personalized icon packs, collections of wallpapers, and even whole themes that change the system's overall look.

To give customers more options, Android also allowed developers to make widgets that could be positioned on the home screen, adding another level of personalization. In addition to giving users access to apps or helpful information, widgets can be designed to blend in with the device's general design. For instance, users may select a bright, colorful widget that goes well with a specific wallpaper or a minimalist design with black and white hues. The need for theming and styling apps has increased due to the ability to integrate widgets into the home screen and change their appearance, enabling users to build genuinely customized and one-of-a-kind mobile experiences.

However, historically, customizing options have been more limited with iOS, the operating system created by Apple for its iPhone and iPad devices. A uniform user experience, which includes a strictly regulated design aesthetic across all devices, has always been Apple's main priority. Consequently, unlike their Android counterparts, iOS users have had fewer possibilities for customizing their smartphones. Previously, iOS only let users alter fundamental components, such as the home screen's app organization and wallpaper. Although users could change the wallpaper to suit their tastes, they had little control over how the system appeared and felt. With iOS 14's introduction of widgets, which increased home screen personalization, this started to alter. More customized devices were made possible by the ability of users to install third-party widgets and organize them in different sizes.

Despite iOS's historically restricted theming capabilities, users could access more customization options because of the growth of jailbreaking, which is the process of eliminating Apple's restrictions. Installing unapproved programs and altering system settings, including changing the look of system components like app icons and fonts, are made possible by jailbreaking an iOS device. Although jailbreaking offered more customization options, there were serious risks involved, such as the possibility of voiding the device's warranty, lowering security, and causing system instability. As iOS continued to enhance its native customization options—especially by adding features like Dark Mode and enhanced widget support—jailbreaking decreased in frequency.

Custom icon-focused apps, aside from launchers, are among the most widely used categories of theming and styling applications. One of the most noticeable visual components of a mobile device's user interface is its icons, and altering their appearance can drastically change how the device looks and feels. Users of icon pack applications can choose from various icons in different styles, from simple and flat to more complex and detailed. A unified visual theme that suits the user's tastes can be produced by substituting these icons for the home screen's default program icons. A user may choose, for instance, a nature-themed icon collection with hand-drawn pictures or an icon pack with pixelated graphics that are retro in style.

Thousands of icon packs, many made by independent designers and developers, are now accessible through app stores as a result of the increased desire for personalized icons. These icon packs frequently include a range of themes and categories to accommodate various interests, and some are free, while others are costly. Icon sets that complement a variety of aesthetics, from minimalist and contemporary to antique and artistic, are thus available to consumers. A great deal of customizing is possible because users frequently combine various icon packs to give their devices a unique appearance.

The wallpaper is another important component of styling and theming. The background images that appear on a mobile device's home screen or lock screen are known as wallpapers, and they are crucial in establishing the device's tone and atmosphere. Wallpaper apps provide users with vast possibilities, such as beautiful artwork, abstract designs, and high-quality images. These applications frequently have enormous image collections that let users look for particular themes or aesthetics, such as architecture, space, or wildlife. To add vitality to the device's appearance, many wallpaper programs now let users set wallpapers that change regularly, automatically, or at predetermined intervals.

Some wallpaper apps offer live wallpapers, which are dynamic or animated backdrops that react to user input via touch gestures or device movements, in addition to static images. By giving the device a more dynamic and interactive feel, live wallpapers can produce captivating and immersive experiences that increase the sensation of customization. When selecting wallpaper alternatives, users frequently need to balance usefulness and aesthetics because live wallpapers can be resource-intensive and may affect the device's performance and battery life.

Typography is an important component of theming and styling. Numerous theming apps enable users to install and apply unique fonts to their devices, even though most mobile devices have preset fonts. An app's or system interface's overall design heavily relies on typography, which affects readability, accessibility, and the user experience's emotional impact. Users can select styles, from professional and elegant to whimsical and informal, by personalizing a device with custom fonts. Some apps even let users change the font's size, weight, and spacing to modify further how text appears on their smartphones.

By providing features that increase the usability of devices for those with visual impairments or other accessibility requirements, theme and style apps also help to promote accessibility. Users with low eyesight or motor abilities can more easily manage their smartphones thanks to the high-contrast color schemes, larger text alternatives, and adjustable layouts in many theming apps. For instance, users can select more readable color schemes or change to a high-contrast setting for better visibility. In this sense, theming and styling applications contribute to a more accessible and inclusive mobile experience for all users and offer visual customization.

Social media have also influenced the popularity of theming and styling applications. Users started posting their customized device configurations on social media sites like Instagram, Twitter, and Reddit as mobile customization gained popularity. These platforms have developed into gathering places for theming communities, where members share advice, discuss ideas, and display their work. These communities have grown even more due to the popularity of these challenges, in which users compete to design the most inventive or aesthetically pleasing sets. Users encourage others to investigate the possibilities of theming and styling apps by sharing their settings, which advances the continuous development of mobile customization.

There are certain difficulties and disadvantages with theming and styling apps, even with their many advantages. The possible effect on performance is one of the main problems. Using launchers, widgets, and live wallpapers to personalize a device's look can use up system resources like memory and processing power, which might result in decreased battery life or slower performance. This is especially true for gadgets with less sophisticated technology. Certain theming apps may also cause compatibility problems because upgrades to the app or operating system may cause some functions to stop working. For instance, a custom launcher may stop functioning after an upgrade to the Android operating system, or an app update may result in misaligned custom icons.

Furthermore, certain theming tools can be too complicated for inexperienced users. While some applications have straightforward and easy-to-use interfaces, others necessitate a more thorough comprehension of system settings and customization options. For instance, users may need to change settings for app placement, gestures, and animations to configure a smartphone using a third-party launcher completely. This could be difficult for people who are not familiar with the procedure. This limits the overall appeal of theming and styling apps by making some users feel overwhelmed or annoyed when modifying their devices.

Despite these obstacles, the need for mobile personalization is still increasing, and theming and style apps continue to play a significant role in the mobile ecosystem. Theming apps may get more complex as mobile operating systems develop and new technologies like machine learning and artificial intelligence are incorporated into mobile devices. AI could be used, for instance, to automatically change a device's look according to the user's preferences or external variables, such as the user's location or the time of day. The capacity of users to build dynamic, individualized mobile experiences may be further improved by such advances, thus solidifying the role of theming and styling apps in the digital world.

# CHAPTER III

# Backend, Databases, and APIs

## Introduction to Backend Development

A crucial component of creating contemporary websites and software applications is backend development. Backend development includes building the systems, servers, databases, and application logic that enable an application's operation, whereas frontend development concentrates on the user-facing portion of the application. The backend is essentially the foundation of an application, and when it is properly designed and functional, it guarantees that the frontend user experience is seamless and effective. Backend development is essential in today's world for building scalable, dependable, and high-performing systems, ranging from e-commerce websites to social media platforms and all points in between.

Backend development is fundamentally about controlling an application's server-side functionality. This entails responding to client requests, maintaining databases, and ensuring all server-side logic and processes run correctly. The infrastructure that supports these features is built and maintained by backend developers. They create the code that allows the application to interact with databases, outside services, and other crucial elements. This code guarantees accurate and effective data processing, storage, and front-end delivery. Backend developers work in the background to ensure an application runs as planned, reacting to user requests and supplying the information and features required for the program to perform its functions.

The function of a server is among the first things to comprehend in backend programming. Servers are strong computers or systems that manage requests from client devices, like mobile apps or web browsers, and store data. Requests are received by a server, which processes them, obtains information from a database if needed, and returns the relevant response. Usually found in data centers, servers can be virtual or physical computers. Servers are configured and maintained by backend developers to make sure they can manage incoming traffic, process requests quickly, and provide clients with data on time. In essence, the server serves as a bridge between the client and the important features of the application, enabling smooth user interaction.

Databases are another essential component of backend development, in addition to servers. Programs use databases to store and manage data, including transaction histories, product details, and user information. Backend developers use databases to generate, retrieve, update, and remove data as the application requires. Although there are many different databases, relational and NoSQL are the most popular. Structured query language (SQL) interacts with data stored in tables in relational databases such as MySQL, PostgreSQL, and SQLite. NoSQL databases, such as Couchbase, Cassandra, and MongoDB, are perfect for managing massive amounts of dynamic data because they store data in a more flexible, unstructured format. The application's particular requirements, such as the type of data, scalability needs, and access speed, determine which database is best.

APIs (Application Programming Interfaces), which facilitate communication between various software components, are another tool backend developers use. A collection of guidelines and procedures known as an API makes it possible for one software program to access the features or information of another. APIs link an application's frontend and server-side functionality when it comes to backend development. Without requiring direct communication with the server, APIs allow front-end applications—whether web-based or mobile—to request and receive data from the backend. GraphQL and REST (Representational State Transfer) are two common architectural techniques for creating APIs. GraphQL enables clients to ask exactly the required data, increasing efficiency and flexibility, whereas REST APIs employ HTTP methods like GET, POST, PUT, and DELETE to interact with resources.

The control of application logic is another key idea in backend development. This code specifies how the backend manages user interactions and data processing. Programming languages like Python, Java, Ruby, PHP, or JavaScript (Node.js) are commonly used to write application logic. Backend developers use these languages to design the procedures and algorithms that power the application's functionality. Application logic could, for instance, manage user authentication, process payments, or carry out intricate data analysis. Usually included in the server-side code, the server executes this logic in response to each request.

Additionally, backend developers must be adept at comprehending and applying permission and authentication procedures. By confirming users' identities, authentication ensures that only people with permission can access particular resources or carry out specific tasks. Numerous techniques can accomplish this, including OAuth, two-factor authentication, and username and password combinations. However, The actions a user can conduct after being authenticated are determined by authorization. For instance, an application might give various users varying degrees of access; for example, common users might be able to read material, while administrators would not be able to change settings or manage data. Authentication and authorization must be implemented correctly for an application to be secure and sensitive user data protected.

Another important factor in backend development is scalability. Applications need to manage higher traffic and data volumes as they expand and draw in more users. The term "scalability" describes an application's backend's capacity to accommodate increasing workloads or user counts without sacrificing efficiency. Numerous strategies can accomplish this, including load balancing, database sharding, and utilizing cloud computing resources. Load balancing divides incoming traffic across several servers to prevent any one server from becoming overloaded. With database sharding, a big database is split into smaller, easier-to-manage chunks housed on a different server. Thanks to the scalable architecture offered by cloud computing platforms like AWS, Google Cloud, and Microsoft Azure, developers may readily add resources as needed to meet demand.

Another crucial backend development method is caching. Caching entails keeping that data in memory to cut down on the time it takes to retrieve frequently accessed material later. Backend systems can save money by caching data instead of executing costly database queries, or API calls each time a user makes the same request. This can greatly enhance an application's responsiveness and performance. There are several layers at which caching can be carried out, including the client side, database, and server. Redis and Memcached are popular caching systems that store key-value pairs in memory for easy access.

Backend developers must also consider security when creating and managing an application's server side. Malicious attacks such as SQL injection, cross-site scripting (XSS), and denial-of-service (DoS) assaults frequently target web applications. Backend developers must use security best practices including input validation, encryption, and appropriate access control to guard against these risks. Input validation guarantees that user-provided data is secure and free of harmful code that could jeopardize the system. Sensitive information, like payment details and passwords, is protected by encryption, transforming it into an unintelligible format only authorized individuals can decipher. Only the resources and data users can see or alter are accessible, thanks to access control methods.

Backend engineers must understand development frameworks to support all of the features above. Frameworks are pre-made collections of tools and libraries that give developers ready-made answers to common problems, allowing them to write code more quickly. Examples of well-known backend frameworks that make web application development easier are Django (Python), Ruby on Rails (Ruby), Laravel (PHP), and Express (Node.js). By offering capabilities like routing, authentication, and database administration, these frameworks free developers from having to reinvent the wheel for fundamental functionalities so they can concentrate on creating the application's distinctive features.

Backend development often involves a series of steps: planning, coding, testing, and deployment. Developers collaborate with stakeholders to specify the needs and features of the application during the planning phase. To illustrate how the backend and frontend will communicate, wireframes or flow diagrams may need to be made. Developers start creating the backend after the requirements are apparent, putting the required database interactions, server-side logic, and API functionality in place. After coding is finished, testing is done to ensure the program works as intended, is error-free, and complies with security requirements. Users can then access the application after it has been deployed to a live server. Following deployment, developers monitor the application's functionality and make any necessary modifications or enhancements.

Backend development, which focuses on the server-side functionality that drives contemporary applications, is an essential component of software engineering. Managing servers, databases, APIs, and application logic and ensuring the application scales effectively fall within the purview of backend developers. Additionally, they must guarantee that the application is safe, dependable, and properly functions in various scenarios. Backend developers may build reliable applications that provide a flawless user experience if they fully grasp backend technologies and strategies, including authentication, caching, and scalability. Backend development is still a vital and dynamic area supporting the technical framework of contemporary digital experiences as web and mobile applications advance.

## Working with Databases

One essential component of contemporary software development is working with databases. Thanks to databases, which efficiently store, manage, and retrieve data, applications can interact with vast amounts of structured or unstructured data. Databases are the foundation of most applications, from enterprise-level systems to mobile apps, whether they are used for handling complicated interactions or for simple data storage. Knowledge of various topics, including database design, database types, data querying, and performance optimization, is necessary to comprehend how to deal with databases. Database management systems (DBMS), data models, database normalization, SQL querying, and crucial factors like data integrity, security, and scalability are all covered in this section's exploration of the fundamentals of database work.

A database is a collection of well-organized data that is simple to manage, change, and access. A database management system (DBMS), which is a collection of software tools that operate as an interface between the user or application and the data contained in the database, is used by developers to communicate with databases. Relational, non-relational, and object-oriented databases are only a few broad categories into which DBMSs can be divided. Every DBMS type has advantages and disadvantages, and the application's requirements will determine which DBMS type is best.

These days, relational databases are the most widely used database type in software development. They use tables to arrange the data; each table has rows and columns. Each row represents a record, and each column represents a field or characteristic inside a record. Since relational databases are founded on the relational model, data can be connected through many-to-many, one-to-many, and one-to-one relationships. MySQL, PostgreSQL, Oracle, and Microsoft SQL Server are well-known relational database management systems (RDBMS) examples. These databases rely on SQL to perform tasks like querying, inserting, updating, and removing data, and they utilize a defined schema to specify the data structure.

NoSQL databases, often known as non-relational databases, are made to manage unstructured or semi-structured data. NoSQL databases don't depend on tables or preset schemas like relational databases. Rather, they employ a variety of data models, including document stores, key-value pairs, column-family stores, and graph databases. Applications that need performance, scalability, and flexibility while handling massive amounts of data that might not fit neatly into a relational structure frequently employ NoSQL databases. Redis, Couchbase, Cassandra, and MongoDB are a few types of NoSQL databases. These databases will benefit from big data, real-time analytics, and applications needing fault tolerance and high availability.

The object-oriented database, which combines database administration with the ideas of object-oriented programming, is another significant kind of database. Instead of storing data as rows and columns, object-oriented databases store it as objects. This method is advantageous for applications that work with complicated data structures because it enables programmers to directly map objects in the programming language to objects in the database. Inheritance, encapsulation, and polymorphism—features common to object-oriented programming languages like Java and C++—are supported by object-oriented databases. Despite being less popular than relational and NoSQL databases, object-oriented databases are used in some application types with intricate data interactions.

Creating an efficient database design is one of the most important aspects of the development process. Knowing what data must be saved and how it will be utilized is one of the first steps in database design. An Entity-Relationship (ER) diagram graphically depicts the system's entities and their relationships, which is the first step in database architecture. Anything or idea that requires data storage, such as a customer, order, or product, is called an entity. The relationship between these entities is defined by terms like "one-to-many" or "many-to-many." For instance, a customer may place many orders in an e-commerce system, and one order may include multiple products. To guarantee that data can be effectively retrieved and changed, the database design must meticulously define these relationships.

Another crucial idea in database architecture is normalization. It describes arranging data to minimize dependencies and redundancies, guaranteeing data efficiency and integrity. Normalization creates relationships between big, complicated tables and divides them into smaller, easier-to-manage ones. First Normal Form (1NF), Second Normal Form (2NF), and Third Normal Form (3NF) are among the normal forms that are followed in the procedure. Repetitive groups must be removed, atomic values must be present in every column, and transitive dependencies must be eliminated, among other requirements for each normal form. Normalization guarantees that the database can be maintained with little effort and helps decrease data abnormalities, such as update, insert, and delete anomalies.

Developers can start working with the data after the database design is finished. Structured Query Language (SQL) is the main language for organizing and querying data in relational databases. SQL allows users to select, insert, update, and delete data, among other database actions. While the INSERT command adds new records to the database, the SELECT statement obtains data from one or more tables. The DELETE statement deletes records from the database, while the UPDATE statement edits already-existing records.

SQL offers more sophisticated functions, such as combining several tables, grouping data, and filtering results using conditions, in addition to the fundamental CRUD (Create, Read, Update, Delete) activities. Developers can use joins to merge data from many tables according to a common column, like an order or customer ID. Depending on how the data must be incorporated, there are several join types, such as INNER JOIN, LEFT JOIN, and RIGHT JOIN, each of which has a distinct use case. Data aggregation benefits from grouping, such as determining the sum, average, or count of specific fields. Only pertinent data is returned using the WHERE clause to filter results according to particular criteria.

Specialized query languages or database-specific APIs are frequently used for accessing NoSQL databases. For instance, Redis employs commands with key-value pairs, but MongoDB uses a query language based on JavaScript syntax. NoSQL databases offer benefits in terms of scalability and flexibility, especially for applications that deal with massive volumes of unstructured data or require real-time data processing. Still, they usually do not enable the same complicated queries as relational databases.

Data integrity is a crucial factor to consider when using databases. It guarantees the accuracy, consistency, and dependability of the data kept in the database. The application of constraints—rules that prohibit the entry of inaccurate or inconsistent data into the database—is one of the fundamental tenets of data integrity. Examples of common constraint types are Primary, foreign, unique, and check constraints. A foreign key creates a connection between two tables, whereas a primary key gives each entry in a database a distinct identity. Check constraints impose particular criteria on column values, such as making sure an age value is greater than zero, whereas unique constraints guarantee that entries in a column are unique.

Another key idea in database administration, especially with relational databases, is transactions. A series of actions handled as a single piece of work is called a transaction. Even in the case of a system breakdown or other disruptions, transactions guarantee that the database stays consistent. A transaction's attributes are defined by its ACID (Atomicity, Consistency, Isolation, Durability) qualities. Atomicity guarantees that a transaction's operations are either successfully performed or not at all. The database's transition between legitimate states is guaranteed by consistency. Durability ensures that its modifications are irreversible once a transaction is committed, while isolation ensures that transactions do not conflict.

Performance becomes a crucial factor when working with databases as applications grow in size and the volume of data increases. Various strategies, including query optimization, caching, and indexing, are used to improve database performance. By enabling the database to find rows based on particular columns swiftly, indexes are data structures that increase the speed of data retrieval. Developers can drastically reduce the time it takes to extract data from big tables by building indexes on frequently searched columns. Caching entails keeping often requested data in memory to minimize the number of database queries needed. This can lessen the strain on the database and is especially helpful for applications requiring quick data access. To ensure that the database can manage high request volumes without experiencing any lag, query optimization entails examining and revising SQL queries to make them more effective.

Another key consideration when working with databases is scalability. The database must accommodate growing loads without experiencing performance issues as the number of users and data volume rise. Vertical and horizontal scaling are the two main methods for growing a database. A single database server's resources can be increased vertically by adding CPU, memory, or storage. Vertical scaling has capacity constraints and can result in a single point of failure despite being comparatively easy to execute. By dividing the data among several servers, horizontal scaling, on the other hand, enables the database to manage increased traffic and storage capacity. Although this method offers greater scalability and fault tolerance, it may be more difficult to install.

Database security is an important consideration in addition to scalability and performance. Maintaining the integrity and privacy of the application requires safeguarding sensitive data against loss, corruption, and unauthorized access. Auditing, authorization, authentication, and encryption are common security procedures. Data is safely kept and sent thanks to encryption, which prevents unwanted people from reading it. While authorization ensures users have the right permissions to access particular data or carry out specific tasks, authentication confirms users' identities. Auditing entails monitoring and recording database activity to identify and address possible security breaches.

In summary, database work is essential to contemporary software development, necessitating knowledge of different database kinds, design concepts, and administration procedures. To guarantee scalability and dependability, developers must be able to create effective data models, query data using the right languages, and improve performance, whether dealing with relational databases, NoSQL databases, or object-oriented databases. Maintaining the accuracy and consistency of the data stored in the database also requires considering factors like data integrity, security, and transaction management. Developers may create reliable apps that satisfy user demands and effectively manage massive data sets by grasping these ideas.

## API Integration and Data Fetching

For applications to communicate with external systems and services and get or transmit data, API integration, and data fetching are essential elements of contemporary software development. Applications must now rely on APIs (Application Programming Interfaces) to operate efficiently and provide dynamic content as the digital ecosystem develops. Applications can interface with other software, platforms, and services through API integration, which increases their capabilities. Concurrently, data fetching allows programs to obtain, process, and present data from various sources. The significance of API integration and data fetching, as well as their underlying procedures, best practices, and the difficulties in properly integrating them into the development lifecycle, will all be covered in this section.

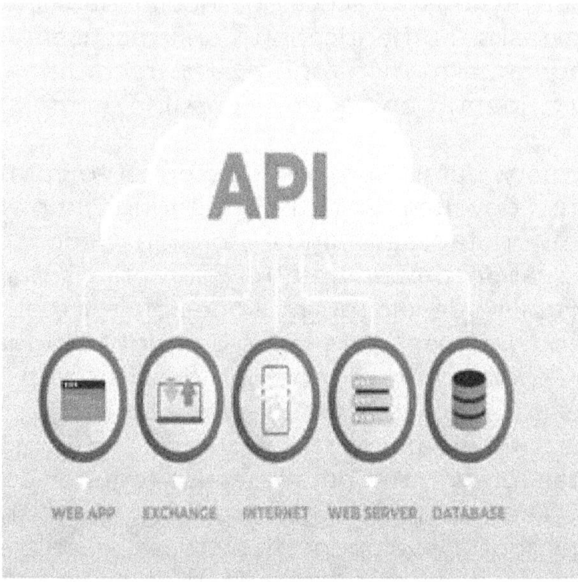


The idea of API integration is to make it possible for various software systems to communicate with one another without any problems. By acting as a conduit for information and functionality between different systems, an API enables programmers to create applications beyond their initial design limitations. Applications can access external services through APIs, such as cloud storage, social media sites, payment gateways, and geolocation services. Developers can save time and effort using pre-built functionality rather than creating everything from scratch when using APIs. For example, a developer can manage login functionality by integrating an existing API service, such as OAuth or Firebase Authentication, instead of developing a whole authentication system.

Connecting an application to a third-party API or service and setting it up to communicate with the external system in a way that satisfies the application's needs are the steps involved in API integration. Usually, this interaction entails sending HTTP queries to the API's endpoints, getting answers, and then processing the information in response. Depending on the use case and the kind of API, request types like GET, POST, PUT, and DELETE may be utilized. These requests relate to tasks like data retrieval, submission, update, or record deletion. Understanding the API documentation, which lists the available endpoints, request forms, authentication methods, and response types, is necessary for the integration process.

The creation of mobile and web applications, as well as enterprise software systems, frequently uses API integration. Finding the right API to integrate based on the required functionality is the first step in the integration process. The next step after choosing is understanding the API's authentication methods. Many APIs need an authentication procedure to provide safe and approved access to the service. This authentication is usually carried out via API keys, OAuth tokens, or other security protocols to verify the identity of the application or user seeking access to the API.

After authentication, developers will interact with the API's endpoints—basically, URLs that represent particular actions or resources within the external system. For instance, a payment gateway API might include an endpoint for handling payments, while a weather API might provide an endpoint for getting the most recent weather information. Specific parameters, like user credentials, data filters, or search queries, must be included in the request for each endpoint. Depending on the API, the request parameters could change. Therefore, developers must ensure they supply the necessary data in the right format. When a request is successful, a response containing the requested data or a status message is often returned as a JSON or XML object.

Developers must manage error scenarios that may occur during the API integration process in addition to making queries and getting answers. Numerous factors, such as network problems, invalid API keys, erroneous queries, or exceeded rate limitations, might cause these errors. Developers should put in place error-handling systems that can deal with these circumstances amicably and give users insightful feedback. For instance, the application should notify the user that the data is unavailable if the API returns a 404 error, which indicates that the requested resource does not exist.

On the other hand, the process of getting data from an API and making it usable by the application is known as data fetching. Sending a request to an API endpoint, getting the data, and then storing or processing it for later use are all steps in this process. Applications that must display dynamic content, such as news feeds, stock market prices, or social network updates, or depend on real-time data must include data fetching. For example, a travel app might retrieve flight information from an airline API, while a news website might utilize a news API to retrieve the most recent headlines.

Developers must consider how data is received and managed within the application to handle data fetching effectively. Using asynchronous programming approaches is a popular strategy that enables the application to send a request to an API without interfering with other processes or the user interface. This is especially significant in contemporary web applications where user experience and responsiveness are essential. Asynchronous approaches allow developers to start data fetching queries in the background and refresh the user interface after the data has been fetched to keep the application fluid and dynamic. Asynchronous data fetching is frequently handled by technologies like AJAX (Asynchronous JavaScript and XML), JavaScript's Fetch API, and libraries like Axios.

Besides asynchronous solutions, the frequency of data fetching and data management after retrieval are additional factors developers need to consider. Since this can result in excessive server load and longer response times, retrieving data from an API for each user interaction is frequently not required. Alternatively, developers can use caching techniques to save retrieved data locally in RAM or a client-side storage system such as IndexedDB or localStorage. Caching enhances overall performance and user experience by enabling the application to offer data faster without requiring repeated API requests. A weather app might save the results of the most recent weather query in its cache and only update the information when the user navigates to a new location or after a predetermined amount of time.

Pagination is another crucial factor to take into account while retrieving data. Many APIs return Large data collections in pages or pieces rather than all at once. By splitting the material into smaller, easier-to-manage chunks, pagination aids in data volume management. Developers must ensure the application can manage and display several data pages when integrating with APIs that allow pagination. To handle edge scenarios like reaching the end of the data set, this may need putting "next" or "previous" buttons in place for page navigation.

Although it is essential to contemporary web and mobile applications, data fetching is difficult. Managing the performance and dependability of API integrations is one of the most frequent problems. API outages may impact the user experience, sluggish response times, or other problems. Developers should use techniques like retries and fallbacks to lessen these risks. Temporary network problems can be resolved using a retry mechanism, which can automatically try a failed request again after a short wait. To ensure that the application keeps running even if the API is momentarily unavailable, fallbacks can also show cached data or a notice stating that the requested data is unavailable.

Managing rate constraints set by certain APIs is another difficulty. To guard against misuse and guarantee equitable resource usage, many APIs limit the number of requests that can be submitted in a specific amount of time. Developers must know these rate constraints and build their applications appropriately because exceeding them may result in temporary or permanent bans. Developers might use techniques like request batching, backoff systems, or dispersing requests over time to prevent reaching rate constraints.

When connecting APIs and retrieving data, security is also a crucial factor. Developers must ensure that the integration process is secure and that any data fetched is handled appropriately because APIs sometimes include sharing sensitive data. This includes encrypting data while it is being transmitted via secure communication routes like HTTPS. Developers should also exercise caution when working with authentication tokens and credentials, ensuring they are kept private and never made public in client-side code.

Data fetching and API integration are critical components of contemporary application development that allow software systems to communicate with other services and retrieve dynamic data. While data fetching concentrates on obtaining and displaying data within the application, API integration entails linking apps to other services. Factors like authentication, request handling, performance optimization, caching, and error management must all be considered during these procedures. The advantages of API integration and effective data fetching are indisputable, enabling developers to design feature-rich, responsive, and data-driven applications—even though security, rate restrictions, and reliability issues must be handled. By grasping these ideas, developers may create robust apps that give users access to pertinent, real-time content, improving their experience in general.

## Authentication and User Management

Modern software systems must incorporate user management and authentication features to guarantee that only authorized users can access resources and sensitive data. These ideas are essential for maintaining the security and integrity of systems, especially as the use of online services and digital platforms increases. Confirming a user's identification, usually using credentials like usernames and passwords, biometric information, or other elements, is known as authentication. To guarantee that users have proper access to the system, user management, on the other hand, includes managing user accounts, permissions, roles, and access controls. Alongside developments in digital security, several technologies approaches, and best practices have emerged in response to the increasing demand for safe and effective user management and authentication systems.

It is impossible to overestimate the significance of authentication, particularly in light of the growing sophistication of cyberattacks. Using a login and password is the most fundamental type of authentication. This approach, essential to many online systems, including email services, social networking platforms, and banking websites, depends on what the customer already knows. Nevertheless, despite being the conventional approach to protecting user accounts, passwords have serious flaws. Attackers frequently infiltrate accounts through phishing attempts, weak passwords, and password reuse. Multi-factor authentication (MFA), which asks users to present many pieces of evidence to confirm their identity, has been incorporated into many systems to overcome these problems. These extra factors could be something the user owns, like a physical security token or a smartphone for a one-time password (OTP), or they could be intrinsic to the user, such as biometric information like voice patterns, fingerprints, or facial recognition.

By guaranteeing that an attacker cannot obtain access without the other necessary elements, even if one factor (such as a password) is compromised, multi-factor authentication significantly boosts security. In a standard MFA configuration, for instance, a user would be asked to enter a code produced by an authenticator app or provided to their phone via SMS after logging in with their password. This two-step procedure significantly lowers the possibility of unwanted entry because attackers would need access to the user's phone or other second factors in addition to their password. MFA is increasingly used, especially in applications that handle sensitive data, like business systems, cloud storage, and banking apps. MFA improves security, but there are drawbacks as well. Using several authentication methods may be cumbersome for users, and some techniques—like SMS-based codes—have been questioned for their efficacy due to their potential for interception. Because they provide stronger protection without the disadvantages of conventional methods, alternative MFA techniques like biometrics and hardware security keys are becoming increasingly popular.

Another crucial area of attention for digital system security is user management. Creating and maintaining user accounts, allocating suitable roles and permissions, and ensuring users can access only the required resources are all part of user management. In addition to metadata about the user's activities and preferences, a user account usually includes identifying information such as the user's name, email address, and password. Creating and overseeing several user roles, such as administrators, moderators, and ordinary users, each with varying degrees of access to system capabilities and data is another aspect of user management for applications with numerous users or organizations. Implementing the principle of least privilege, which states that users should only be given the minimal amount of access required to do their jobs, is the aim of user management.

User administration can get increasingly complicated in larger systems, especially in enterprise settings. Organizations may need to manage thousands or even millions of user accounts with different access levels across numerous platforms and applications. Centralized identity and access management (IAM) systems are frequently used. With the help of these technologies, businesses can regulate user access and authentication across several systems from a single location. Administrators can create and implement access policies using technologies like role-based access control (RBAC), fine-grained permissions, and single sign-on (SSO) commonly found in IAM systems. SSO enhances security and user ease by enabling users to authenticate only once and access all linked systems without entering their login information again. RBAC makes it simpler to manage user access based on job duties by allowing administrators to assign users to roles, each with unique rights.

Modern user management systems frequently include functions like user provisioning and de-provisioning, which automate generating and removing user accounts, in addition to these fundamental capabilities. The user management system, for instance, can automatically create an account for a new hire, assign them to the proper job, and provide them with access to the resources they need when they join the company. On the other hand, the system can immediately remove an employee's account and deactivate their access when they depart the company, guaranteeing that no illegal access is left behind. In regulated sectors like finance and healthcare, where user access needs to be closely monitored and managed, this procedure is essential for preserving the security and compliance of an organization's systems.

Monitoring and auditing are crucial components of user management. Security-conscious enterprises use continuous monitoring systems to monitor user activity and identify any odd activity that could point to a security breach. For instance, the system might raise an alarm for more inquiry if a user accesses a significant amount of sensitive data or logs in from an unknown location. Additionally, auditing systems can capture comprehensive user action logs, which are very helpful in determining the underlying cause of security problems or guaranteeing adherence to industry standards. Regular audits of user accounts and permissions are required to ensure that only authorized users have access to important resources and that any unused or out-of-date accounts are quickly terminated.

Handling the intricacy of managing identities across many platforms and systems is one of the difficulties related to user management. An example would be a user having several accounts in several programs, such as social media, cloud storage, and email. If the user must have several usernames and passwords for every account, managing these divergent identities can be time-consuming and prone to mistakes. Identity federation, a technique that enables users to authenticate once with a central identity provider and then access many services without having to log in again, was developed in response to this issue. Federated identity solutions enable users to easily access many apps while preserving a single identity across systems by utilizing protocols like Security Assertion Markup Language (SAML) and OpenID Connect.

The growth of cloud computing and mobile applications has brought about additional user management and authentication considerations. Managing resource access across a distributed infrastructure is becoming more and more difficult for businesses as they move to cloud-based platforms. Several third-party services and a combination of public and private resources are frequently used in cloud environments, further complicating user access security. Organizations are using multi-cloud identity management solutions more regularly to handle these issues. These solutions enable them to secure and control user identities and rights across various cloud providers and services. By integrating with current IAM systems, these solutions allow administrators to apply uniform access controls to all cloud resources.

Authentication and user management have also changed due to mobile applications, commonplace in daily life. Securing mobile access is difficult since mobile devices are naturally portable and frequently utilized in various settings. In contrast to conventional password-based authentication, mobile authentication solutions usually use biometrics, such as fingerprint scanning and facial recognition, which provide convenience and increased security. MFA is also supported by many mobile apps, enhancing their protection against unwanted access. Organizations may enforce security standards across fleets of mobile devices with mobile device management (MDM) systems, which also secure access to individual applications. Sensitive data is safeguarded even in cases where devices are compromised thanks to MDM systems, which enable managers to remotely change settings, impose encryption, and remove data from misplaced or stolen devices.

The regulatory environment for user management and authentication has tightened due to the increased focus on privacy and data protection. Stricter guidelines for the gathering, storing, and processing of personal data have been imposed by laws like the California Consumer Privacy Act (CCPA) in the US and the General Data

Protection Regulation (GDPR) in the EU. According to these rules, businesses must have robust authentication systems, keep user consent procedures open and transparent, and give users access to and control their personal information. Companies must make sure that their user management procedures adhere to these rules as part of this, which may entail putting features like user rights management, data encryption, and safe data storage into place.

The ever-changing nature of cybersecurity threats always challenges systems for user management and authentication. Organizations must remain alert and proactive in implementing new technology and best practices as fraudsters create increasingly complex ways to get around established security systems. This entails implementing cutting-edge authentication techniques like biometric recognition, behavioral biometrics (which examine user behavior patterns to spot irregularities), and artificial intelligence (AI)-driven threat detection systems, which can spot possible security threats before they become serious problems. AI has enormous potential to enhance security and usability when incorporated into user management and authentication systems. AI systems, for instance, can automatically modify security settings based on variables like the user's location, device kind, and previous behavior to make authentication as easy and safe as feasible.

User management and authentication are essential to contemporary digital system security. The tools and procedures used to verify users and control their access to resources must change along with the digital environment. The methods and tools to protect systems from unwanted access are cloud-based identity management, identity federation, user roles and permissions, and multi-factor authentication. To maintain the strength and efficacy of their user management and authentication procedures, companies must keep innovating and adapting in the face of the constant threat of cyberattacks. Ultimately, the objective is to balance security and usability, allowing users to safely access the resources they require while reducing the inconvenience and friction that excessively complicated authentication procedures may cause.

# CHAPTER IV

# Implementing Core App Features

## Handling User Input and Forms

Managing forms and user input is an essential part of web development since it directly affects how users engage with an application or website. Forms frequently collect user data, including payment information, comments, and registration details. In addition to functionality, user experience and security depend on how well user input is managed in these forms. The procedure entails gathering data, verifying it, and then processing or storing it in a way that guarantees precision, security, and usability.

Any data entered into a system by a user is called user input. This is usually done through various form fields, such as text boxes, checkboxes, dropdown menus, and radio buttons. Applications ranging from user registration for a service to e-commerce checkouts depend on this data. Developing an appealing, user-friendly interface that motivates users to enter accurate data is the first hurdle in managing user input. This is the point at which the form's design becomes crucial. An enjoyable user experience is influenced by the design, the instructions' clarity, and the error-handling procedures. Users may stop the process before finishing a form if they find it difficult to use or comprehend. As a result, web developers concentrate on making easy forms to use, ensuring that every field is well-labeled and that the form makes sense.

Validating the input to ensure it satisfies the requirements comes after a user submits the form. Client-side validation and server-side validation are the two categories of validation. Client-side validation occurs in the user's browser before the form is sent to the server. JavaScript is usually used, which can check for missing fields, improper formats, or even basic errors like a phone number that does not fit the needed pattern. Client-side validation gives the user instant feedback, which improves the user experience by enabling them to fix mistakes without waiting for a server response.

Client-side validation is insufficient on its own, though. Users can get around it despite its benefits by purposefully changing the form data or disabling JavaScript. Server-side validation is essential for this reason. Once the form has been sent to the server, server-side validation occurs. At this point, the server double-checks all the data to ensure it satisfies the standards, including having a secure password, correct credit card information, or an email format. Malicious users trying to take advantage of form flaws, including inserting malicious scripts or evading client-side validation rules, are prevented by server-side validation.

Preventing malicious attacks like SQL injection, cross-site scripting (XSS), and cross-site request forgery (CSRF) is one of the most crucial parts of safely processing user input. These assaults use system flaws to obtain unauthorized access or perform destructive tasks. To access or alter the database, malicious SQL code is inserted into a form field. This technique is known as SQL injection. By separating user input from the SQL query and using prepared statements and parameterized queries, developers can avoid this and ensure that the input is handled as data rather than executable code. In XSS attacks, malicious scripts are injected into input fields and run in a different user's browser. Sensitive data, such as login passwords or personal information, may be stolen. Sanitizing and escaping user input and eliminating or neutralizing any potentially dangerous scripts before they are processed or shown on the website are all part of preventing cross-site scripting attacks (XSS).

CSRF, another frequent security issue, deceives users into submitting a form against their will. Developers employ anti-CSRF tokens, which are distinct values added to forms that confirm the validity of the request, to protect against CSRF attacks. The request is denied if the token does not match the expected value upon submission of the form. Developers can guarantee the safe handling of user data and the security of the form submission process by fixing these security flaws.

Helpful error messages must be provided in addition to security when managing user input. Users should be informed in a straightforward and welcoming way when they make a mistake, whether a missing field or an erroneous format. Error messages describe what went wrong and offer instructions on how to fix it. When a user inputs an invalid email address, the error message may read, "Please enter a valid email address in the format user@example.com." To avoid confusing users, error messages should be placed next to the impacted fields and should not be extremely complex or cryptic. Using both client-side and server-side validation to provide immediate feedback, users may immediately correct their errors without repeatedly submitting the entire form.

Additionally, forms must be made to accept different user input. Considering accessibility elements for individuals with disabilities is part of this. For example, form fields should be appropriately labeled with clear instructions, and users who cannot use a mouse should be able to navigate the form using a keyboard. Developers should make sure that they use semantic HTML and ARIA (Accessible Rich Internet Applications) roles to make the forms accessible because screen readers need to understand them. Additionally, giving users a range of input choices—like date pickers for calendar entries or auto-complete for address fields—can simplify the form-filling process.

Processing or storing the data comes next after the input has been verified. In certain instances, the information—like user profiles or order details—will be kept in a database for later use. To protect sensitive data, such as passwords or credit card numbers, it is crucial to maintain this data securely. Where appropriate, encryption and hashing techniques are used. For instance, developers can keep a hashed version of a user's password instead of the password itself. This one-way function makes it harder for attackers to determine the password's reverse. Developers should also abide by data privacy laws like the General Data Protection Regulation (GDPR) to guarantee that personal data is handled appropriately and lawfully.

To sum up, managing user input and forms requires several essential components, including designing intuitive user interfaces, validating data, guaranteeing security, offering constructive criticism, and processing data effectively. A proper implementation guarantees the security and integrity of the data gathered and enhances the user experience. To ensure that the forms operate efficiently and securely, developers must focus on every aspect, from form design to error handling and security protocols. The ability to handle user input has become crucial for any web developer due to the rise in online transactions and the significance of user data.

## Accessing Device Features

Since most modern gadgets offer many functionalities beyond basic communication, accessing device features has become crucial to modern technology. The capabilities built into smartphones, tablets, laptops, and wearable technology allow users to do various functions, such as work, communication, entertainment, and even health management. Usually, using the hardware and software components of the device is necessary to access these features. Operating systems, programs, user interfaces, and occasionally direct hardware controls help ease this connection. How consumers access these functions changes as devices become more intelligent and connected, offering a more seamless and richer experience.

Access to more sophisticated features is made possible by the fundamental parts of any device, such as the processor, memory, and sensors. For instance, combining a photo sensor, processing power, and specialized software is necessary for a smartphone to take high-resolution pictures. Usually, users interact with built-in software, such as the native camera app, or launch apps to access these functionalities. The development of machine learning and artificial intelligence (AI) is making device features more predictive and intuitive. As a result of these technologies' ability to learn from user behavior, accessing device functionalities becomes more customized.

When giving users access to gadget functionalities, operating systems (OS) are essential. They manage resources and make sure features function properly by serving as a bridge between the user and the device's hardware. Operating systems like Windows, macOS, iOS, and Android are made to have intuitive user interfaces that make it simple for people to access a wide range of functionality. For instance, a smartphone's operating system (OS) uses a variety of touch gestures, voice commands, and app interfaces to provide users access to the camera, microphone, sensors, GPS, and many other functionalities. Without an operating system, most users would find handling the low-level instructions needed to access these functionalities difficult or impossible.

Numerous sensors built into contemporary electronics allow access to functionalities based on input from the surroundings. Accelerometers, gyroscopes, barometers, proximity sensors, and ambient light sensors are some examples of these sensors. For example, a user may use a heart rate sensor to track health metrics or an accelerometer to track movement when using a fitness app. Location-based services that assist users in navigating or checking in to locations in real-time may be made possible via the phone's GPS technology. Likewise, voice-activated functions, such as virtual assistants like Alexa, Google Assistant, or Siri, can be accessed through the microphone. These sensors offer useful information that improves the user experience and enables the gadget to carry out sophisticated tasks.

Voice interfaces have made it much easier to access the functions of devices. Many contemporary gadgets come with voice assistants that let consumers complete activities without using their hands. Voice assistants use a device's microphone and software to process requests, such as inquiring about the weather, sending a message, setting a reminder, or managing smart home appliances. These assistants' strength is their natural language comprehension, which enables users to easily access capabilities without navigating menus or interfaces. Additionally, people with physical limitations or others who find it difficult to engage in traditional modes of interaction will benefit most from voice-based access to device functionalities.

Touch interfaces, in addition to voice control, have completely changed how people use gadgets. Thanks to the development of capacitive touchscreens, users may now access device functionalities with just a tap, swipe, or pinch gesture. Touch interfaces make Multitouch functionality possible, enabling users to complete several tasks simultaneously. These operations could be rotating an image, scrolling over a webpage, or zooming in on a map. Users now have even more control over the device's functionality because of touchscreens' increasing sensitivity and ability to recognize new gestures like force touch and haptic feedback. One of the primary reasons smartphones and tablets have become such essential components of daily life is the ease with which touch interfaces can access functionality.

The Internet of Things (IoT) further complicates how we access gadget functions. Nowadays, many gadgets have internet connections and cooperate to offer more functionality and ease. For instance, a smart home system might enable customers to use voice commands or a mobile app to manage the lights, thermostats, or security cameras. Wireless technologies, including Wi-Fi, Bluetooth, and Zigbee, are used by Internet of Things devices to connect. Users can use these technologies at home or on the go to manage and monitor features from almost anywhere. Users can now access device functions from the device itself and via remote or cloud-based systems thanks to the growth of the Internet of Things.

Access to device functions has also changed due to the cloud, especially regarding data processing and storage. Cloud services make accessing data from many devices easier, enabling users to save files, images, music, and other stuff remotely. Cloud-based applications can improve performance by offloading computationally demanding activities or granting access to more resources than a single device could manage independently. For example, a device might use a cloud service to carry out intricate machine learning or data analysis operations that would be impossible to complete locally because of hardware constraints. Furthermore, cloud-based syncing services improve accessibility and user experience by guaranteeing that device functionality like contacts, calendar entries, and files remain the same across several devices.

Accessing gadget features raises serious security and privacy issues, especially as devices become increasingly networked and can handle sensitive data. Biometric authentication technologies, such as fingerprint scanning and facial recognition, have gained widespread acceptance as safe ways to access devices and their functionalities. Thanks to these authentication techniques, only authorized users can access certain device functionalities, such as banking apps or personal data. Encryption is used to safeguard data while it is being transmitted, especially when using cloud services or the internet to access device functionalities. Although the level of protection offered by these security measures is better, there are difficulties in balancing privacy concerns and convenience. Users must take precautions to protect their privacy and be aware of the security dangers while using specific functions on gadgets that are increasingly capable of gathering personal data.

Last but not least, the way individuals use technology has changed significantly due to the advancement of device functions. Accessing a device's functionality is no longer passive as it gets smarter and more user-friendly. Through app creation and third-party integrations, users may now alter settings, personalize their devices, and even add new capabilities. As a result, developers are producing apps that increase the capability of devices in previously unthinkable ways, sparking an explosion of creativity. The ability to access device functionalities has become a fundamental aspect of the human experience, and as technology develops, there are countless opportunities for new and enhanced device interaction.

To sum up, using a device's features is a constantly changing process that combines user interaction, software, and hardware. How we access and use the features of contemporary gadgets is influenced by several factors, including operating systems, sensors, voice assistants, touch interfaces, the Internet of Things, and cloud services. Accessing these capabilities will become easier as technology develops, providing customers with more convenience, utility, and customization options. At the same time, protecting user privacy and security will remain a top priority while developing new technologies. Ultimately, the capacity to effortlessly and intuitively access device functionalities has revolutionized our way of life, work, and communication, influencing technology for years.

## Adding Multimedia and Animations

A fundamental component of contemporary web and app design is using animations and multimedia in digital platforms. Due to technological advancements, users now anticipate dynamic, captivating, and immersive experiences when interacting with digital information. To provide visually appealing content, multimedia—which includes images, videos, audio, and other interactive elements—has become an essential tool. On the other hand, animations enhance the user experience by adding movement and interactivity to digital items. Multimedia and animations are incorporated into digital products to boost usefulness and aesthetic appeal. This makes interfaces easier to use and guides users through activities or processes. The significance, guiding ideas, and difficulties of incorporating multimedia and animations into digital information will be discussed in this section, along with how they affect user engagement and experience.

Using several content kinds, including text, images, audio, and video, on a digital platform is called multimedia. By providing a variety of interaction options, multimedia integration into a website, application, or game can greatly improve the overall experience. For instance, websites frequently employ images to enhance page aesthetics, break up content, and clarify ideas. To provide customers with a good picture of the products they are considering buying, e-commerce platforms must have high-quality product photos. Images can reinforce a company's or product's identity as part of a branding plan.

Another potent type of multimedia that may make an otherwise static interface active is video content. The popularity of social media networks that facilitate video sharing and streaming services like YouTube and Vimeo has made people more used to watching videos online. Users can access instructional ads, behind-the-scenes videos, and any other content that would benefit from a moving visual by integrating video into a website or app. A fitness app might incorporate workout videos to walk users through various activities. To manage the playback experience, videos can be hosted locally on a website or server or embedded via third-party services. Either way, adding video makes content delivery more interesting than text or still photos.

Audio is another multimedia element that has various uses in digital design. Audio can be utilized for voiceovers, sound effects, and background music on websites and applications. In gaming, audio is essential because it enhances images and provides sound effects that react to player movements, creating an immersive experience. Similarly, audio is frequently employed in educational apps to support learning through narrations or voice instructions. However, excessive audio usage can result in a disruptive user experience; thus, it must be used carefully. For instance, background music that automatically plays can irritate people attempting to concentrate on reading or interacting with other page elements. Maintaining a satisfying user experience in these situations requires having obvious controls to mute or change the volume.

A primary advantage of incorporating multimedia into digital platforms is the capacity to accommodate various user requirements and preferences. While some people could benefit more from auditory content, others prefer visual stuff like pictures and films. Multiple engagement channels made possible by multimedia guarantee that users with different learning preferences or styles can interact with the content, whatever best suits them. For example, visual learners and those who prefer written instructions may benefit from video training paired with text-based instructions. Multimedia can also increase accessibility for people with disabilities. Users with hearing or vision impairments can access content more easily when movies have subtitles, or photos have written descriptions.

Similar to multimedia, animations are essential for producing dynamic and captivating experiences. Fundamentally, animations are the movement of components in a digital domain; they are frequently employed to highlight particular regions of a website or screen, produce transitions, or give visual feedback. It is impossible to overestimate the significance of animations in interface design. They can direct users, convey modifications, and improve the intuitiveness of interactions. Navigation systems are one area where animation is frequently used. Animated transitions can show consumers that their actions have been recorded when they click a button or link. For example, a button may animate to visually "press down" when a user clicks on it, providing them with instant confirmation that their action has been recognized. This improves the user experience overall by giving a sense of confirmation and responsiveness.

A website or app's content flow and transitions between screens or sections can also be enhanced with animations. For instance, an animation can smoothly transition the information from one page to the next when you navigate between them. In addition to making the digital product feel more professional and user-friendly, this produces a smooth and polished experience. This method is frequently called "micro interactions," which are tiny, nuanced animations that show the user how their actions impact the interface.

Animations can be utilized for teaching as well as for bettering navigation. Animated lessons can help customers better grasp how to use the interface while leading them through a procedure. To help users understand where to click and what to do, banking software might employ animation to demonstrate how to set up a payment or transfer money. The application becomes more user-friendly and accessible by using animations to explain the procedure, which lessens the cognitive burden for users who might otherwise find new activities difficult.

In addition to its practical uses, animations are important for fostering personality and interaction on digital platforms. For example, the interactive features of a social media app or the light-hearted animations of a mobile game can support the brand's identity and create a more pleasurable user experience. Users can develop an emotional bond with a product and become more engaged by animations that complement its characteristics. For instance, a website for a store selling toys for kids might have vibrant, colorful animations that enhance the site's fun and excitement. These animations have the power to produce an unforgettable experience that entices visitors to visit the website or application again in the future.

Although animations and multimedia improve user experience, they also pose some difficulties for designers and developers. One of the primary considerations is ensuring that adding multimedia material doesn't impair the application's speed or performance. Multimedia files can greatly increase a webpage or application's size, particularly high-resolution photos or videos, which might cause longer load times. Users may become irate and abandon the platform due to slow loading times. To solve this, developers frequently employ lazy loading, video compression, and image optimization to guarantee that multimedia material is provided effectively without sacrificing performance. Multimedia elements' file format and resolution should also be taken into account by designers because they can affect loading speed.

Ensuring that animations improve the user experience rather than detract from it presents another difficulty. Users may become frustrated by a disorganized or confusing experience caused by excessive or badly done animations. Animations should be utilized carefully and sparingly to prevent this, ensuring they have a clear function and don't overload the user. Additionally, as choppy or delayed animations can produce an uncomfortable experience, animations should be made to be fluid and smooth. When using animations, accessibility factors also need to be taken into mind. Certain animations may cause discomfort or confusion for some users, especially motion-sensitive users. Designers can include settings that enable users to disable or minimize animations to accommodate these people.

Additionally, animations and multimedia must work with various screens, browsers, and devices. Multimedia components must be responsive and adjust to multiple screen resolutions because people access websites and apps on different devices. This could entail creating layouts with mobile devices in mind, utilizing scalable vector graphics (SVGs) for images, or putting media queries into place to change the size of multimedia components according to the device. Animations should also be tuned to work well across various devices, including tablets, smartphones, and desktop PCs. To ensure that the material works with all major browsers, developers must consider how different browsers support multimedia features.

In summary, animations and multimedia in digital products are crucial for producing visually appealing, interactive, and captivating experiences. Images, videos, and audio are multimedia components that enhance the content by providing consumers multiple ways to interact with the platform. Conversely, animations improve navigation, increase engagement, and offer visual feedback, which makes the interface seem more responsive and user-friendly. However, including animations and multimedia in a product takes considerable preparation and thought, especially regarding accessibility, user experience, and performance. Designers and developers may produce digital experiences that are not only useful but also memorable and pleasurable by carefully combining these components. Multimedia and animations, when applied skillfully, can enhance a digital platform and transform a straightforward user interaction into an interesting and captivating experience.

## Performance Optimization

Performance optimization is a crucial component of software development to increase an application's responsiveness and efficiency while consuming the fewest resources possible. Regardless of their platform or device, people want the software to be quick, responsive, and effective in today's fast-paced digital environment. Performance optimization is essential to guaranteeing a satisfying user experience and optimizing the longevity and usefulness of software, whether it is desktop, mobile, or web-based. Performance issues include sluggish loading speeds, unreliable user interfaces, and excessive resource usage can lead to disgruntled consumers, lower retention rates, and unfavorable reviews. Therefore, any development team must comprehend the basic ideas of performance optimization and use practical methods to deal with bottlenecks and inefficiencies.

The requirement to increase software's speed and responsiveness is at the heart of performance improvement. The time it takes for an application to react to user input or finish a job is one of the most basic elements influencing performance. Applications like video games, financial software, and communication tools that demand real-time or nearly real-time performance may be crucial. Frustration and discontent can result from slow response times, making a program feel unresponsive and sluggish, particularly when engaging with the user interface. Reducing the time it takes the system to process input and provide results is a common aspect of performance optimization. This can be accomplished in several ways, such as resource management, code optimization, and algorithm enhancements.

Finding performance bottlenecks is one of the first steps in performance improvement. A bottleneck is a system component that restricts an application's overall speed or throughput. The application's user interface, network communication, data processing, and storage access are just a few locations where these bottlenecks may appear. Because optimizing the incorrect part of the program can result in only modest gains in overall performance, locating the bottleneck is essential. Profiling tools and performance analysis approaches are crucial to diagnosing performance problems and identifying areas that require improvement. For instance, web developers can find issues with resource utilization, JavaScript execution, and page load speeds using tools like Lighthouse or Google Chrome's Developer Tools. Similar to this, developers of mobile apps may keep an eye on CPU, memory, and network activities by using tools like Android Profiler or Xcode Instruments.

Developers can solve the problem by using a variety of optimization strategies after the bottleneck has been located. For instance, reducing the time it takes for the user interface to update in response to user input is frequently the aim of UI optimization, which aims to make the program more responsive. This can frequently be accomplished using strategies like lazy loading, which loads content just when necessary rather than all at once. This enhances initial load times and keeps the system from becoming overloaded. Reducing the rendering process's complexity is another method for improving UI performance. For instance, complex animations, transitions, and visual effects can significantly tax the system's resources and result in subpar performance, particularly on devices with lesser specifications. Reactivity can be increased by simplifying animations, minimizing the number of components on the screen, or shifting some work to background threads.

Algorithms and data structures can be optimized at the code level to increase performance in addition to UI changes. In many situations, an algorithm's efficiency can significantly affect performance. For instance, a sorting algorithm that is not well-optimized may take a long time to run, particularly when dealing with huge datasets. This could lead to delays and higher resource use. Developers can lower the application's temporal complexity and speed up execution times by selecting more effective algorithms, like quicksort or merge sort. Similarly, performance and memory use can be significantly impacted by choosing the right data structures for a certain activity. For example, the application's speed and efficiency can be maximized by employing an array for sequential data or a hash map for quick lookups.

Another essential component of speed optimization is memory management. In severe situations, slow performance, high memory utilization, and crashes can result from inefficient memory usage. Over time, performance issues may also be caused by memory leaks, in which the system fails to release memory that is no longer required. Developers must carefully control memory allocation and deallocation to prevent memory leaks and excessive usage. In languages like C and C++ that lack integrated garbage collection, this may entail manually maintaining memory or utilizing automatic memory management methods like garbage collection. For instance, optimizing the memory utilization of mobile apps is crucial to ensuring that they function properly, even on devices with constrained capabilities. Effective memory management is essential for avoiding slowdowns and crashes since mobile devices often have less memory and processing capability than desktop PCs.

Network communication is another area where performance enhancement is being done. Many contemporary applications depend on network connections to access remote data, services, and APIs. However, network communication can cause serious performance problems when dealing with sluggish or unreliable connections. Developers can use techniques like asynchronous data fetching, which lets the application run. At the same time, it waits for data to arrive from the network, or data compression, which lowers the amount of data that must be sent over the network to maximize network performance. In web development, methods like content delivery networks

(CDNs) and HTTP caching help speed up the distribution of material to end users and lessen the strain on servers. Images, scripts, and stylesheets are frequently used resources that can be cached by the application to prevent repeated calls to the server, lowering latency and enhancing performance.

Performance can be enhanced by decreasing the quantity of requests sent to the server and improving network connection. For instance, developers can utilize methods like GraphQL to request only the necessary data or aggregate requests into a single batch request instead of making many individual requests for each piece of data. In addition to improving performance, fewer requests also lessen the strain on the server, preventing server overload and enhancing scalability. Improving communication between components can significantly affect overall performance in distributed systems, where applications depend on numerous services and databases. The system can scale efficiently and manage large traffic volumes without incurring performance degradation with the use of strategies like load balancing and service orchestration.

Another crucial component of speed optimization is database optimization, especially for applications that store and retrieve a lot of data. Because ineffective queries or inadequately indexed data can result in delayed response times and heavy server loads, databases are frequently the bottleneck in many applications. Developers can employ strategies like query optimization, which involves rewriting inefficient SQL queries to improve performance, and indexing, which involves indexing frequently visited data to facilitate quicker searches and retrieval to maximize database performance. Developers can also use caching techniques to speed up data retrieval and minimize database requests. Caching enables the application to swiftly retrieve frequently used data without repeatedly querying the database by storing it in memory.

The underlying technology and infrastructure should also be considered while optimizing database performance. For instance, data retrieval times can be shortened by utilizing quicker storage options like solid-state drives (SSDs). By distributing the load among several servers or storage devices, distributed databases and sharding can also increase scalability and performance. To guarantee that the system can manage the anticipated traffic and load without needless overprovisioning, performance optimization in cloud-based systems frequently entails choosing the appropriate cloud instance types and allocating resources optimally.

Performance can be impacted by how software is distributed and maintained in addition to optimizing an application's technical features. Continuous integration (CI) and continuous delivery (CD), two deployment methodologies, might guarantee the consistent and dependable application of performance enhancements. Development teams can ensure that optimizations are not inadvertently missed or neglected by automating the testing and deployment of performance enhancements. Additionally, as the application develops, continuous monitoring and performance tracking are crucial for locating new bottlenecks and places needing improvement. Application performance management (APM) software is one example of a performance monitoring tool that can give developers important information about how an application functions in real-world settings and help them identify problems before they affect users.

Optimizing mobile applications is one subject that has attracted much interest lately. Performance optimization is particularly difficult for mobile devices since they are frequently limited in memory, computing power, and battery life. Mobile developers must optimize their apps' influence on battery life, resource utilization, speed, and responsiveness. Techniques like effective power management and background task optimization, for instance, can assist in reducing battery drain, and fewer background operations can enhance system performance. Additionally, mobile apps must be optimized for various device requirements because smartphones and tablets differ significantly in terms of screen size, processor speed, and memory capacity.

To sum up, performance optimization is a crucial component of software development that directly influences resource usage, system effectiveness, and user experience. Developers can use various tools and strategies to improve application performance, such as optimizing database queries, lowering network latency, enhancing memory management, or optimizing algorithms. Performance optimization will continue to be a key area of study as software systems increase in complexity and scale, necessitating constant testing, analysis, and improvement. Developers can produce software that satisfies customers' growing needs while guaranteeing effective and efficient use of resources by adopting performance optimization best practices and utilizing the newest tools and technologies.

# CHAPTER V

# Testing and Debugging

## Debugging Common Issues

Any software developer must be able to debug typical problems. Since no code is ever completely free from mistakes or unexpected behavior, it is an unavoidable step in development. The process of locating, separating, and resolving issues in software is known as debugging. Simple syntax errors, more intricate logical errors, or runtime issues that arise under particular circumstances might all be included in this category. Debugging is essential because it guarantees that the program operates effectively, fulfills user expectations, and functions as intended. Effective debugging is necessary for developers to produce dependable, high-caliber code.

Syntax errors are among the most frequent problems developers face. These are errors in the code that go against the programming language's grammatical rules. Since the compiler or interpreter usually offers a clear error message indicating the line of code where the issue occurred, syntax mistakes are generally easy to discover. Missing semicolons, improper parentheses, unclosed strings, and misspelled keywords are examples of common syntax mistakes. While fixing these problems is simple, they can still be annoying, particularly if the notice is ambiguous or the error is concealed within a lengthy code block. Examining the code thoroughly, looking for mistakes, and confirming that all syntax rules are followed to debug syntax issues is critical. The built-in linters or syntax checkers found in many text editors and integrated development environments (IDEs) can assist in identifying these mistakes before executing the code.

Logic mistakes are another common problem that developers encounter. These mistakes happen when the code executes without crashing but yields inaccurate or surprising outcomes. Since logic errors don't result in software failure, they are more difficult to spot than syntax faults. Rather, they result in unexpected behavior or inaccurate output. For instance, a developer may design a function that determines a shopping cart's total cost, but a missed mathematical operator could cause the calculation to be incorrect. The logic and flow of the code must be carefully examined to debug logic issues. Adding print statements or using a debugger to track the program's execution and verify variable values at various points is one useful method. Developers can identify where the reasoning goes wrong by knowing how the program arrives at its faulty conclusion.

Runtime problems are another frequent problem in debugging. These happen when the software runs and frequently cause the application to crash or behave strangely. Invalid input, such as trying to divide by zero, accessing an array element that isn't there, or executing a function with the wrong arguments, is usually the source of runtime errors. Because they rely on particular circumstances or inputs that might not always exist during development or testing, these problems are frequently more difficult to debug. Developers can employ various techniques to troubleshoot runtime issues, including logging, exception handling, and step-by-step debugging. Logging helps document the program's state and determine the events leading up to the issue. By catching exceptions and keeping the program from crashing, exception handling can assist in successfully managing problems. Finding the error's location and cause is easier using debugging tools that let you look through the code and look at the stack trace.

Another frequent problem is memory leaks, especially in programming languages without automatic garbage collection, such as C and C++. When a program allocates memory for data or objects but neglects to release it when it is no longer required, this is known as a memory leak. These memory leaks can build up over time, impairing system performance and ultimately causing crashes. Developers must utilize specialized memory profiling tools to monitor memory utilization and pinpoint spots where memory is being allocated but not released to debug memory leaks. By ensuring that memory is appropriately deallocated when it is no longer required, developers may address the problem by using these techniques to identify the locations of the leaks. Memory leaks can still happen in garbage collection languages like Python or Java because references are kept in use longer than necessary, making it impossible for the garbage collector to recover the memory. In these situations, examining object references is critical and ensuring superfluous ones are removed.

Debugging concurrency problems in multithreaded programs is especially difficult. It becomes challenging to predict the sequence of execution when several threads or processes are operating concurrently, which can result in deadlocks, race situations, and other synchronization issues. Unpredictable outcomes might arise when two or more threads try to alter shared data simultaneously, a problem known as a race condition. It's called a deadlock when two or more threads are stuck waiting for each other to release resources. A thorough grasp of thread interaction and a close look at how shared resources are accessed is essential for debugging concurrency problems. By monitoring the sequence of thread execution and resource access, tools such as thread analyzers, profilers, and logging can assist developers in locating deadlocks or race circumstances. Since concurrency issues frequently depend on timing and might be difficult to reproduce in regular operation, producing them in a controlled environment can also be beneficial.

Dealing with external dependencies, such as third-party libraries, services, or APIs, presents another debugging difficulty. Errors may arise if the software depends on external components that aren't working correctly or have been updated in a way that conflicts with the code. Because the developer may not have control over the problem's origin, these problems can be particularly difficult to troubleshoot. For instance, server outages or modifications to the API's response format could cause a call to an external API to fail. Debugging in these situations starts with determining whether the problem is with the code or the external dependence by looking at the external service's status.

Additionally, developers should review the documentation for any modifications to the library or API that might have impacted their code. Recording external service responses and faults can give important information about the problem's location. Additionally, during debugging, interactions with external dependencies can be simulated using mock services or stubs.

When debugging, developers need to be aware of security flaws in addition to these typical problems. Because security flaws like SQL injection, cross-site scripting (XSS), or improper authentication frequently entail intricate interactions between the code and outside parties like attackers, they can be challenging to debug. Understanding secure coding techniques and the possible dangers of incorrectly processing user input or external requests are essential for debugging security flaws. Code review is a popular method where developers look for potential security vulnerabilities in the code. Using automated security testing tools that can look for common vulnerabilities in the application is another beneficial technique. Additionally, logging can be used to spot unauthorized access attempts or questionable activities that might point to a security problem.

An essential step in the debugging process is testing. Before they ever get to the debugging stage, numerous problems can be identified via automated tests, including unit, integration, and end-to-end tests. While integration tests examine how various components interact with one another, unit tests verify that individual parts or functions operate as intended. End-to-end tests confirm that the entire program functions as intended by simulating real-world usage scenarios. Automated testing gives developers a safety net by promptly finding and addressing problems they may otherwise overlook. Continuous integration (CI) tools can also automatically run tests whenever code is posted to the repository to ensure that new changes don't bring new bugs.

To sum up, debugging is a crucial talent for software developers and requires technical expertise, patience, and problem-solving skills. Runtime issues, concurrency problems, syntax errors, and logic faults are just a few of the many problems developers commonly face. Ensuring the application runs correctly and consistently is the universal objective, albeit each problem calls for a different strategy. Tools like debuggers, profilers, and logging systems facilitate debugging and offer important insights into the application's behavior. Developers can enhance their debugging abilities and create more reliable, error-free software by adhering to best practices and learning to recognize typical problems.

## Testing Your App

Testing your app is one of the most important stages of the app development process. Ensuring the application works as planned is crucial, offering a flawless user experience while upholding strict security and performance guidelines is vital. Testing is essential for finding and resolving bugs, enhancing usability, and confirming the app's functionality. In order to find problems early and fix them effectively, testing should be included in the development cycle rather than just taking place after the program has been constructed. The many testing methodologies used in app development, available tools and frameworks, and best practices for efficient testing are all covered in this section.

Testing is to confirm that the application satisfies the stated requirements and operates as intended in various settings. This entails testing the application for errors, inconsistencies, and usability problems while ensuring it functions best on different hardware, operating systems, and network configurations. Testing is crucial for identifying bugs, but it confirms that the application offers a positive user experience, complies with industry norms, and operates dependably in practical settings. The testing procedure can be divided into multiple phases, each concentrating on the application's distinct aspects, including compatibility, security, performance, and functionality.

Unit testing is among the primary testing to consider while creating an application. Unit testing focuses on confirming the app's separate parts or features to ensure that every code section operates as intended. Developers create these tests, which are usually automated, to examine particular features or procedures separately from the rest of the program. Because they assist in identifying problems early in the development process, unit tests are crucial in preventing minor errors from later developing into more significant concerns. Automated unit testing frameworks, like Mocha for JavaScript, XCTest for iOS, or JUnit for Java, are frequently used to streamline and improve the testing process. Because they offer a safety net to identify mistakes during code updates and refactoring, unit tests are especially helpful when working with complicated codebases.

Integration testing, which confirms that various app components function as intended, is another crucial kind of testing. Integration tests ensure these parts work correctly when combined, whereas unit tests look at individual modules or functions separately. Integration testing is essential when several modules or services must communicate with one another, like when an application wants to retrieve data from a server or access device APIs. For instance, integration tests may be used to confirm that, when the payment gateway is integrated with the app's shopping cart system, the checkout procedure functions as intended while testing an e-commerce application. This kind of testing aids in identifying problems that occur when various app components depend on one another. Espresso for Android, Postman for API testing, and UI Automator for testing the interactions between UI elements in Android apps are popular frameworks for integration testing.

Another crucial component of app testing is user interface (UI) testing. UI testing guarantees the application's interface is aesthetically pleasing, useful, and easy to use. It ensures users can browse the app without problems by testing its responsiveness, layout, and design aspects on various screen sizes and devices. UI testing for mobile apps must also consider different screen sizes, landscape and portrait device orientations, and other peculiarities unique to the platform. Both automated and manual UI tests are possible. To replicate user interactions and confirm that the application reacts appropriately to user input, including button clicks, text input, and swiping motions, automated UI testing frameworks like Selenium, Appium, or XCUITest are frequently utilized. On the other hand, manual UI testing uses actual users engaging with the application to find any usability, layout, or design faults. Understanding the whole user experience and ensuring the app satisfies its target audience's expectations require this type of testing.

Another crucial aspect is performance testing, which examines how effectively the application functions in various scenarios. Performance testing checks the app's speed, stability, and responsiveness to ensure it functions properly, even in scenarios requiring a lot of resources or heavy traffic. Performance testing measures several important characteristics, including load time, memory utilization, CPU usage, and network latency. For example, an app with excessive memory usage or slow loading times may result in a bad user experience and high abandonment rates. Performance testing includes load testing, stress testing, and scalability testing. Load testing mimics high-user traffic to find out how well an application manages a high number of concurrent users. Stress testing looks at the application's performance under harsh scenarios, including a high server load or restricted bandwidth. Testing for scalability ensures the application can grow effectively as the volume of data or user base grows. Firebase Performance Monitoring, LoadRunner, and Apache JMeter are well-known performance testing solutions.

Security testing is essential to ensure the app is safe from potential threats, data breaches, and vulnerabilities. Security testing is becoming a crucial step in app development due to the rise in cyberattacks and data protection concerns. The goal of security testing is to find potential flaws in the database, network connectivity, or app code that hostile actors could use. Weak encryption, unsafe API endpoints, incorrect handling of sensitive data, and authentication errors are examples of common security problems. Ethical hacking, or penetration testing, is frequently carried out to mimic possible assaults and find weaknesses before they can be taken advantage of. Verifying that the application complies with privacy laws, such as the California Consumer Privacy Act (CCPA) or the General Data Protection Regulation (GDPR), which mandate particular safeguards for user data, is another aspect of security testing. To find security holes and vulnerabilities in web and mobile applications, utilize tools like AppScan, Burp Suite, and OWASP ZAP.

Making sure the app is user-friendly, intuitive, and fits the demands of the intended audience is the main goal of usability testing. Usability testing assesses how well users can engage with the app, as opposed to functional testing, which verifies that the app functions as intended. It entails seeing actual users engage with the software to spot issues, unclear workflows, or components that might not be immediately apparent. The app's design, navigation, and general user experience are all validated with the use of usability testing. Usability specialists frequently use this testing to evaluate the app's usability, accessibility, and user happiness. User input from the real world is crucial for enhancing the app's user interface and ensuring that it meets the needs of the intended user base.

Cross-platform testing is also crucial to ensure the software works correctly across many platforms and devices. Thanks to the widespread usage of smartphones, tablets, and wearable technology, users now engage with apps on various devices, operating systems, and screen sizes. Cross-platform testing guarantees that the application functions uniformly on several platforms, including the web, iOS, and Android. To ensure the app works as intended, it must be tested across various devices, operating systems, and browser versions. Because it guarantees that users on multiple devices have the same experience, this testing is especially crucial for apps that want to reach a large audience. Tools like TestComplete, Sauce Labs, and BrowserStack frequently facilitate cross-platform testing.

Compatibility testing verifies that an application functions on diverse hardware configurations and operating system versions, another crucial factor to consider while testing an application. Testing the application on previous versions is essential to guarantee backward compatibility because operating systems are frequently changed. Variations in device hardware specs, such as processing speeds, RAM, and camera quality, may also impact the app's performance. Making sure the app works as intended on all target devices and operating system versions is made easier with compatibility testing.

Regression testing must be done once the testing phase is over before the app is released. Regression testing ensures that bug patches or new features don't unintentionally interfere with already-existing functionality. This testing confirms that, even after code modifications, the application continues to function as intended. To verify that everything in the app still works as intended, a series of pre-defined tests can be run using automated testing frameworks like Selenium or Appium. Regression testing guarantees that the program stays stable and dependable by identifying flaws that might have gone unnoticed in previous stages of development.

Acceptance testing is the last phase of testing, which confirms that the application satisfies the demands and guidelines established by stakeholders, including clients, project managers, and end users. Usually, acceptance testing is done right before the software is put into production. It entails confirming that every feature is available, the application runs as planned, and all performance, security, and usability requirements are satisfied. This kind of testing is the last stage before the app is made available to users or uploaded to the app store, and it may include both automated and manual tests.

To summarize, testing your app is a crucial step in the development process that helps guarantee its usability, security, performance, and usefulness. Unit testing, integration testing, user interface testing, performance testing, security testing, usability testing, and cross-platform testing are some of the stages that make up the testing process. To find and fix problems early in the development process, utilizing the right testing tools and frameworks and adhering to best practices is crucial. Developers can guarantee a top-notch user experience, lower the possibility of bugs and vulnerabilities, and enhance the app's overall efficiency by extensively testing it before releasing it.

## Performance Testing and Optimization

In the software development lifecycle, performance testing and optimization are essential procedures that guarantee that applications function properly in various scenarios and satisfy customer demands for dependability, speed, and responsiveness. Performance testing aims to identify bottlenecks or inefficiencies that can impair system performance by assessing how well a system performs under various workloads and circumstances. Contrarily, optimization entails improving the system's speed, responsiveness, and scalability to boost performance. This procedure is essential for apps that manage high data or traffic volumes and guarantee end users a smooth and pleasurable experience. This section examines the main ideas, methods, difficulties, and best practices related to performance testing and optimization and how they aid in creating high-performing applications.

A subset of non-functional testing, performance testing, assesses how an application behaves in different performance scenarios. The main objective of performance testing is to ensure that an application can manage the anticipated user load and function effectively in typical, high-volume, and stressful scenarios. It also looks for performance-related problems, like sluggish reaction times, excessive resource usage, or system crashes. Various tests, including load, stress, scalability, and endurance tests, are included in the vast field of performance testing. The purpose of each test type is to evaluate various facets of an application's functionality.

One of the most popular types of performance testing is load testing. To ascertain how well the system can handle the load, it entails modeling the anticipated volume of users or transactions an application will manage under typical circumstances. For instance, load testing can assist in establishing whether a website can withstand increased visits without crashing or becoming slow during special events or promotions. To determine the maximum number of concurrent users or requests the system can manage, load testing usually entails progressively raising the load until it reaches its peak performance capacity.

Another crucial kind of performance testing is stress testing, which assesses how effectively an application functions in harsh environments and beyond its typical capabilities. Stress testing aims to find the system's breaking point, the point at which performance deteriorates, or the point at which the system breaks completely. By conducting stress tests, developers can better understand how an application responds to higher-than-normal demands, like an unexpected rise in traffic or a large influx of data. It assists in locating systemic weaknesses that might not be noticeable in normal circumstances but could become problematic amid unforeseen spikes in traffic or stressful situations.

The goal of scalability testing is to assess an application's scalability—its capacity to manage a growing demand over time. Businesses may need to scale their apps to handle more users or larger transaction volumes as they expand. To satisfy increasing demand, scalability testing helps identify whether the system can scale vertically (by adding additional computers or servers) or horizontally (by increasing the resources of current machines). The aim of scalability testing is to find any restrictions in the infrastructure or architecture of the system that can hinder its capacity to scale efficiently. An application may not be able to manage a rise in user traffic without experiencing noticeable performance loss, for instance, if it mostly depends on a single server.

Endurance testing, sometimes called soak testing, evaluates the system's long-term performance by running the application under a constant load for a prolonged time. This kind of test aids in locating problems that could arise with extended use, such as memory leaks, resource depletion, and performance degradation. Endurance testing guarantees the program can withstand prolonged workloads without malfunctioning by mimicking long-term usage. For example, a cloud-based service that is supposed to function around the clock should undergo endurance testing to ensure that there are no performance problems when used continuously for long periods.

Optimizing the application to achieve optimal performance under a variety of scenarios is the next step once performance testing is finished. Finding and fixing performance bottlenecks, increasing system effectiveness, and making sure the application can smoothly manage the anticipated load are all part of optimization. Performance optimization can be achieved in various ways, and the particular strategy used will rely on the application's characteristics and the performance problems found during testing. The main objectives are improving reaction times, reducing resource usage, and guaranteeing system scalability.

The program's code is one of the main areas that needs improvement. Slow execution times, high CPU usage, and excessive memory consumption can all be caused by inefficient or badly written code. Refactoring inefficient algorithms, cutting down on pointless actions, and enhancing data structures are all part of code optimization. For instance, the system load can be decreased if an application loops through big datasets inefficiently or makes pointless database queries. Developers can concentrate on optimizing the system components that have the most effects on performance by using code profiling tools to find performance hotspots in the code.

Database performance is another crucial area that has to be optimized. Databases are frequently the foundation of applications, and their performance can greatly impact the speed and responsiveness of the system as a whole. Enhancing performance requires optimizing database queries. This includes ensuring that indexes are used efficiently to speed up data retrieval, decreasing the number of database joins, and optimizing SQL queries to reduce complexity. Database normalization and denormalization techniques can also guarantee effective data storage and retrieval. Additionally, caching frequently accessed data helps speed up response times and lessen database strain.

Web servers and application servers are also essential to performance. While application servers conduct business logic and oversee database transactions, web servers respond to incoming requests and transmit messages to clients. Optimizing server setups, enhancing load balancing, and lowering server-side processing overhead are all examples of server-side performance optimization. By dividing up incoming traffic among several servers, load balancing keeps any one server from becoming overburdened. Developers can enhance the system's overall performance by appropriately scaling servers and optimizing server resources like CPU and memory.

Optimizing the front end is just as crucial as optimizing the server-side components. The user interface and the elements that users interact with, including HTML, CSS, JavaScript, and pictures, are the front end of a web application. To decrease the time it takes for a website to load and become interactive, front-end performance optimization aims to reduce the quantity and complexity of the browser's assets. Front-end performance can be enhanced with lazy loading of resources, picture compression, and minification of JavaScript and CSS files. To minimize initial load times, lazy loading ensures that just the required resources are loaded first, with more resources being loaded when the user interacts with the page.

Application performance is significantly impacted by network performance as well. An application's speed can be greatly affected by latency or the delay in data transmission between the client and the server. Reducing the quantity of data transfers, decreasing the number of requests made between the client and the server, and utilizing strategies like content delivery networks (CDNs) to serve material from servers nearer to the user are all ways to improve network performance. CDNs may lower latency and speed up response times by reducing the distance between the user and the server.

Memory management is another crucial component of performance optimization. Memory leaks, unnecessary trash collection, and general performance loss can result from inefficient memory utilization. Developers may maximize memory use by making sure that memory is allocated and released effectively, reducing the creation of superfluous objects, and utilizing memory management strategies like object pooling. Using memory profiling tools, developers can resolve memory leaks and inefficient memory utilization before they affect performance.

After an application has been deployed, it is crucial to regularly monitor and assess its performance. Testing and optimization are continuous activities. Tools for real-time monitoring can assist in spotting performance problems in production so that developers can take proactive measures to fix them. Developers can monitor the application's performance and make necessary modifications by gathering information on response times, resource utilization, error rates, and other performance indicators. Performance monitoring can also assist in spotting user behavior trends that need more improvement, like bottlenecks during high traffic or device performance deterioration.

Despite the abundance of performance testing and optimization techniques, developers still have several obstacles to overcome to guarantee peak performance. One of the biggest hurdles is finding a balance between performance and other goals, such as functionality, security, and maintainability. Excessive optimization attempts might result in complicated, difficult-to-maintain code, even while optimization can greatly increase performance. Developers must balance performance optimization with preserving the system's general adaptability and ease of upkeep. Furthermore, it's frequently challenging to reproduce performance problems in testing settings, particularly in real-world scenarios like different network speeds, hardware setups, and user behavior. It is crucial to test the application in various scenarios to find any issues.

To sum up, performance testing and optimization are essential to guarantee that apps fulfill customer demands for dependability, quickness, and responsiveness. While optimization entails improving the system to increase efficiency and scalability, performance testing assists in locating possible bottlenecks and problems that can impair an application's capacity to manage the anticipated load. Developers can greatly enhance an application's performance by concentrating on code optimization, database performance, server optimization, and front-end optimization. Maintaining peak performance in a production setting requires constant testing and monitoring, and resolving performance problems as they appear contributes to a responsive and seamless user experience. Continuous procedures like performance testing and optimization are essential to creating and maintaining superior apps with outstanding performance.

## Security Best Practices

Security best practices are crucial rules for applications, systems, and data to be safe from possible threats. Adhering to appropriate security procedures is essential for safeguarding sensitive data, guaranteeing business continuity, and upholding user trust in the digital age, where data breaches, cyberattacks, and other malevolent actions are becoming more frequent. Every facet of technology development necessitates reducing vulnerabilities and safeguarding assets against unwanted access, manipulation, or theft, from protecting web applications and network infrastructures to implementing secure coding techniques and conducting frequent security audits. Businesses, organizations, and developers must always proactively identify and resolve security issues to maintain a comprehensive security plan. They also need to stay ahead of new threats.

Adhering to the least privilege principle is one of the main security best practices. This idea centers on granting users, procedures, and systems the minimal access required to perform their duties. Organizations can greatly lessen the harm that compromised accounts or systems could create by limiting access to important resources and capabilities. The least privilege principle applies to system access and user authorization. For example, sensitive data should only be accessible to people who require it for their jobs, and only authorized individuals should have administrative privileges. Applying the least privilege can also lessen the possibility of privilege escalation attacks, which allow an attacker to access higher levels of a system's functionality without authorization.

Enforcing robust authentication procedures is essential for system security and restricting user access. Despite being widely used, password-based authentication is frequently insufficient because of the risks linked to weak or reused passwords. Multi-factor authentication (MFA) should be commonplace to reduce these threats. To confirm their identity, MFA requires users to present several pieces of proof, such as a password, a physical token or mobile device, or biometric information like fingerprints or facial recognition. By adding an extra layer of protection, MFA makes it much harder for hackers to obtain illegal access, even in cases when they have successfully cracked a password.

Regular password management procedures should also be implemented, such as mandating that users change their passwords regularly and ensuring they adhere to complexity standards. This helps keep users from using easy-to-guess passwords susceptible to dictionary and brute force assaults. Furthermore, requiring distinct passwords for every account or service lowers the possibility of credential stuffing attacks, in which hackers combine stolen username-password combinations to gain access to numerous accounts.

Using encryption to safeguard sensitive data is another essential security best practice. Data transmission and storage are guaranteed unreadable by unauthorized parties, thanks to encryption. Encryption helps ensure that even if data is intercepted, it cannot be readily deciphered, regardless of whether it is in transit (sent over a network) or at rest (kept on a disk or server). For instance, safeguarding data sent over the internet requires using Secure Sockets Layer (SSL) or Transport Layer Security (TLS), its successor. By encrypting data sent back and forth between web browsers and servers, these protocols prevent eavesdropping on sensitive information such as financial transactions and user credentials. Furthermore, using robust encryption algorithms for backups and data storage guarantees that data is safe and difficult to exploit, even in the case of a breach or theft.

In addition to encryption, network and communication security is essential for preventing attacks like denial-of-service (DoS) and man-in-the-middle (MITM) assaults. Patching and updating network hardware regularly, such as intrusion detection/prevention systems, routers, and firewalls, helps guard against security flaws that hackers might exploit. Another crucial procedure for safeguarding private information when accessed via public networks is using Virtual Private Networks (VPNs) and other secure communication methods. By separating sensitive systems from less sensitive ones, network segmentation can also lessen the harm a security compromise can do. This approach improves overall security by making it harder for attackers to travel laterally within the network.

Updating systems and software is crucial in guaranteeing that discovered vulnerabilities are promptly fixed. Software developers and vendors often release updates and security patches to fix recently found vulnerabilities. Systems may become vulnerable to possible exploitation if these upgrades are not applied. An automated patch management system can ensure consistent and timely application of updates. Maintaining system security also requires routinely checking software providers' security warnings and searching for missing patches. Furthermore, since third-party dependencies may also have vulnerabilities that need to be fixed, upgrading them is equally as crucial as updating core software.

Writing secure code is crucial to software development to avoid security holes an attacker could exploit. Using secure coding techniques means that software is designed and written with security in mind from the start. Input validation is one essential procedure that guarantees all user inputs are thoroughly examined and cleaned before the application processes them. Inadequate input validation can result in vulnerabilities like buffer overflows, SQL injection, and cross-site scripting (XSS), which can be used to obtain unauthorized access, run arbitrary code, or steal data. Additionally, developers should embrace secure coding practices, which include avoiding hardcoded secrets, interacting with databases via parameterized queries, and routinely checking and testing their code for security flaws.

Another essential security procedure is code review. Multiple developers reviewing each other's code lessens the possibility that security problems will be overlooked. Additionally, automated static code analysis tools can find common security flaws and give developers real-time feedback while working on a project. This method lowers the possibility that vulnerabilities may find their way into production environments by identifying possible security issues early.

Conducting routine security testing is just as crucial as developing secure code. A range of techniques, including risk assessments, penetration testing, and vulnerability scanning, should be used in security testing. Through comparison with a database of common security weaknesses, vulnerability scanning technologies can assist in identifying known vulnerabilities within a system or application. In contrast, penetration testing simulates an attack on the system to find vulnerabilities that attackers might exploit. Organizations can prioritize their efforts to eliminate vulnerabilities, identify possible hazards, and assess their overall security posture with regular risk assessments.

An often overlooked but essential part of a thorough security strategy is implementing an incident response plan. Even with the strongest security measures, assaults and breaches can still happen. Organizations may respond quickly and efficiently to control and lessen the harm caused by a security breach if they have a well-defined incident response plan. Clear processes for locating, documenting, and handling security issues should be part of an incident response strategy and communication guidelines for alerting relevant parties, authorities, and impacted parties. The plan should also specify how to recover from the breach, including restoring systems from backups, conducting a post-incident study, and taking corrective action to avoid similar events in the future.

Another essential part of security best practices is monitoring and auditing. Anomalies, possible security incidents, and new threats can be quickly identified using ongoing system, network, and application monitoring. Systems for security information and event management (SIEM) compile and examine logs from several sources, giving security teams important information about system activity and enabling them to react quickly to suspicious activity. Conducting routine security audits, which involve examining user access logs, system configurations, and security rules, helps guarantee that security protocols are being adhered to and that systems continue to meet industry standards and requirements.

Additional best practices in cloud security must be followed to safeguard data and apps housed on cloud platforms. Although cloud service providers provide several security measures, it is ultimately the responsibility of enterprises to secure their networks, data, and applications. Setting up cloud services effectively, including enabling and configuring encryption, access limits, and monitoring, is one of the main best practices for cloud security. Organizations should also use robust authentication methods like multi-factor authentication (MFA) to secure access to cloud resources. In-transit and at-rest data should be secured, and sensitive workloads should be separated from less important ones using appropriate network segmentation.

Developers of mobile applications must concentrate on protecting the data they manage and the application code. Threats, including data leaks, reverse engineering, and illegal access, are especially dangerous for mobile apps. While data encryption and secure storage techniques help protect sensitive data saved on mobile devices, code obfuscation can make it more difficult for attackers to decompile or analyze the application code. Additionally, developers should use secure communication protocols like HTTPS to safeguard data sent between the server and the mobile app. Additionally, mobile apps should have safeguards against typical vulnerabilities like code injection, session hijacking, and unsafe API calls.

Lastly, a key component of any security strategy is training staff members and users on security best practices. Phishing and other social engineering assaults are becoming more widespread and can be very successful if people are unaware of the dangers. Frequent awareness and training campaigns can assist users in identifying dubious links, emails, and other harmful activity. Additionally, users should be urged to develop sound security practices, such as creating strong, one-of-a-kind passwords for every site and exercising caution when disclosing personal information online.

To sum up, security best practices are essential for shielding data, systems, and apps from the dangers of the modern digital environment. Organizations can greatly lower the risk of security breaches and guarantee the security of their systems by implementing various tactics, including enforcing strong authentication, encrypting sensitive data, adhering to secure coding practices, patching software regularly, and keeping an eye out for suspicious activity. To protect against changing threats and keep users' and customers' trust, a proactive security posture must be combined with regular training, testing, and incident response planning.

# CHAPTER VI

# Launching and Maintaining Your App

## Preparing for App Store and Google Play Submission

To make their apps available to a worldwide audience, developers must go through the crucial and frequently painstaking process of preparing for submission to the App Store and Google Play. To ensure their apps are approved and satisfy the requirements, developers must adhere to certain rules and specifications on both platforms. These specifications cover various topics, including performance, privacy, security, and app functionality. A seamless submission and approval procedure can be ensured by knowing these rules and preparing to prevent delays or rejections.

Ensuring the app is completely functional and free of bugs is one of the first steps in getting ready to submit. Apps must work as promised and properly on various devices to be approved by the App Store and Google Play. To ensure that all features function as intended and that there are no crashes or mistakes, developers must carefully test their apps. In addition to testing the app's core features, testing should cover edge cases and scenarios that could reveal hidden issues. To guarantee compatibility, apps should be tested across various device models, OS versions, and screen sizes. Through thorough testing, developers can avoid frequent problems resulting in rejection throughout the review process.

Following the design standards established by Google and Apple is the next important step. Because design directly impacts the user experience, both platforms place a high value on it. Google employs Material Design principles, while Apple adheres to its Human Interface Guidelines. To give consumers a straightforward and aesthetically pleasing experience, developers must ensure that their app's user interface complies with certain rules. Unfavorable reviews may hamper an app's prospects of success if it seems haphazard or challenging to use. Developers can produce an app that satisfies clearance requirements and provides consumers with a polished and intuitive experience by adhering to platform-specific design principles.

Other important factors to consider during the submission process are security and privacy. There are strict guidelines at Google Play and the App Store to safeguard user information and ensure apps don't violate privacy. To protect sensitive data, developers must utilize strong security methods, including data encryption and secure user authentication. Developers must also include explicit privacy policies that outline the procedures for gathering, using, and sharing user data. The user must be notified whenever the app gathers personal information, and express approval is needed in many situations. Developers must ensure their software satisfies all legal and regulatory standards before submitting it, as noncompliance with privacy and security regulations may lead to rejection or removal from app stores.

Making excellent app assets is yet another crucial step in the submission procedure. To display their app, developers must submit various assets, such as images, promotional videos, and the app symbol. These resources are crucial for drawing users in and providing them with an overview of the features and appearance of the app. While the images should offer a clear view of the app's user interface and highlight important features, the app icon should be straightforward, unique, and symbolic of the program's main objective. The screenshots must adhere to certain dimensions and format specifications for various devices to be published on Google Play and the App Store. Even though they are optional, promotional movies can aid in offering a captivating sneak peek at the software and promoting downloads. High-quality assets are crucial to make a good first impression and raise the app's visibility.

Additionally, developers must include a thorough description when submitting their apps to the App Store or Google Play. The app's features, functions, and user benefits should be spelled out in detail in the description. To increase the app's exposure in search results, developers should utilize clear text that emphasizes the app's unique selling qualities and include pertinent keywords. Since they give information about the features and advantages of the app, effective app descriptions are essential for drawing users in. Developers must provide the app's category, target user base, and any pricing or subscription information in addition to the description. Giving customers thorough and precise information improves the app's discoverability in app stores and aids in their decision-making.

Lastly, developers must ensure their software meets the submission standards particular to each platform. There are specific guidelines for using third-party services, handling payments, and moderating content in the App Store and Google Play. For instance, the App Store mandates using Apple's in-app purchase system for any digital content or subscriptions within the app. Similarly, Google Play mandates that apps adhere to the Play Store's in-app purchase payment guidelines. Developers should ensure their software conforms with all applicable laws and regulations by carefully reading the instructions supplied by both platforms. By doing this, companies can guarantee that their app won't be rejected for technical or policy reasons and prevent delays in the approval process.

Developers can submit their app for assessment after completing all required processes. Depending on the platform and app complexity, the review process usually takes several days to a week. The app will be made accessible on Google Play or the App Store if it is accepted. Developers can fix the problems before resubmitting the app, even if denied. They will receive feedback explaining the reasons for rejection. Following approval, developers must monitor customer input and release updates to fix issues, enhance functionality, or add new features.

In conclusion, meticulous attention to detail and adherence to platform-specific requirements are necessary while getting ready to submit to the App Store and Google Play. Before submitting, developers must ensure their software is secure, well-designed, fully functional, and compatible with privacy laws. Successful approval also depends on accurate descriptions, high-quality app assets, and compliance with submission guidelines. Developers can boost their chances of approval and guarantee that their app reaches a large audience by carefully planning it and taking all required requirements. A successful software launch and user distribution depend on the submission procedure, even though it can be difficult.

## Publishing Your App

An essential phase in the app development lifecycle is publishing your app. It is the last stage before your app is made accessible to people worldwide, whether through the Apple App Store, Google Play Store, or other channels. This phase entails several actions, such as getting the app ready for submission, ensuring it complies with store policies, and handling any technical, legal, and marketing issues. An organized strategy is necessary for a successful app release to guarantee that the application is usable, discoverable, safe, and complies with app store platform standards.

Ensure your software works as intended across various devices and operating systems by extensively testing it before submitting it for publication. The program should have no performance problems, crashes, or defects. Additionally, developers must confirm that their software conforms to the app shops' requirements. The Apple App Store and Google Play have specific app guidelines regarding operation, security, privacy, and content. For instance, Google's developer regulations, which include data collection, user privacy, and forbidden content, must be followed by apps available in the Google Play Store. Similarly, Apple has its own rules, the App Store Review Guidelines, emphasizing design, security, and quality. Becoming familiar with these guidelines before submitting your app is crucial, as noncompliance may lead to its rejection.

The app must then be ready for submission after testing is finished. Developers must create app materials, including icons, promotional films, screenshots, and a clear description, for both iOS and Android apps. These resources are crucial because they give prospective users a sneak peek at the features and UI of the application. An app's chances of being found and downloaded can be greatly increased with the help of eye-catching screenshots and an interesting description. To assist the app show up in pertinent searches inside app stores, developers should also include information, such as the app's name, category, and keywords. The description should inform and entice potential users by highlighting the app's salient features and functionalities.

Developers must also set up any legal or business requirements that may be required when getting ready for publication. For instance, developers need to ensure they have the right licensing for any third-party libraries or software they utilize in their apps. Privacy laws such as the California Consumer Privacy Act (CCPA) and the General Data Protection Regulation (GDPR) must be followed if the app gathers personal information or deals with sensitive data. Before collecting data, developers must get user approval and publish privacy standards on iOS and Android platforms. Apps may occasionally need age limitations, especially if they include inappropriate material for certain audiences. To prevent problems during the app evaluation, developers must set these parameters throughout the submission process.

Developer accounts must be created to publish the program on the Apple Program Store or Google Play Store. Developers must create a Google Play Console account and pay a one-time charge to publish an app on the Google Play Store. To access the Apple App Store, developers must pay an annual subscription fee to join the Apple Developer Program. The app can be uploaded to the platform for evaluation when the developer account has been created. Both platforms have a review procedure to ensure apps adhere to its rules. The review procedure for Google Play is usually quicker, taking only a few days. However, depending on the app's complexity and compliance with Apple's stringent requirements, the approval process for the Apple App Store may take longer —typically a few days to a week. App reviewers will assess the app's performance, security, usability, and functionality throughout the review process. The app will be rejected if any problems are found, and the developer will have to fix them before submitting it again.

Developers ought to consider marketing plans when the app is being reviewed to advertise it after it is released. A marketing strategy must be developed to guarantee that the software reaches its intended audience. Effective tactics to increase the app's visibility include developing a website, setting up social media accounts, and utilizing search engine optimization (SEO). To improve the app's chances of appearing in pertinent search results, developers can also use app store optimization (ASO) by enhancing the app's description, keywords, and images.

Public relations initiatives can help spark interest and awareness, such as asking tech bloggers or influencers to evaluate the app. Starting a paid advertising campaign on sites like Facebook or Google Ads can increase exposure and encourage user growth.

After your app is published, continuing support and upgrades are another important factor to consider. After your app goes live, users could find bugs, ask for additional features, or offer suggestions for improving it. Developers ought to have a plan for responding to customer feedback, providing timely updates, and resolving any issues brought to their attention. Users can offer reviews and ratings on Google Play and the Apple App Store, and developers should keep an eye on these remarks and reply to them. The software may be greatly improved using user input, which can also point out areas needing improvement or optimization. Frequent updates fix compatibility problems with new devices or operating system versions, enhance the app's functionality, and keep users interested.

After the app is released, it's crucial to monitor its performance. This entails monitoring key performance indicators (KPIs) such as in-app purchases, user retention, and downloads. The Apple App Store and Google Play offer analytics tools that let developers know how well their apps are doing. These tools can assist in determining user patterns, trends, and areas needing development. To increase user retention, developers may need to modify an app's functionality, user interface, or marketing strategies if it is experiencing a high uninstallation rate.

Releasing your app is a thrilling but difficult process requiring thorough preparation and close attention to detail. A successful launch depends on ensuring your app complies with platform requirements, creating eye-catching assets, and adhering to privacy and regulatory requirements. After the app is out, continued marketing initiatives, user interaction, and upgrades will support its success. Developers may increase the likelihood that their app will be found, downloaded, and used by people all over the world by adhering to best practices for app release.

## Monetization and Marketing Strategies

Any firm must have monetization and marketing strategies to succeed, but they are especially important in the digital sphere, where intense competition exists and constantly changing consumer behavior. Marketing strategies are the plans and techniques used to draw in and keep customers, whereas monetization is making money from a product or service. These two components work together to create the basis for transforming a concept or product into a successful business. The survival and expansion of a firm can be greatly impacted by knowing the best marketing techniques and the many monetization approaches.

The target market, product, and business type all influence monetization tactics. There are a few standard methods for making money for websites, applications, and digital platforms. Direct sales are the most conventional approach in which customers pay for access to a good or service. This strategy can benefit businesses that provide valuable items like software, media, or online courses. Offering premium content or services worth the price is crucial in this business strategy since it depends on drawing in paying clients.

The freemium model, which charges consumers for access to premium services or content but offers a basic version of the product for free, is another well-liked monetization technique. This strategy works effectively for online platforms, games, and apps when providing a free version draws a lot of users, some of whom become paying clients. Finding the ideal mix between free and paid services is essential to the freemium model's success. While having too many free features can make it hard to persuade people to upgrade, having too few features in the free version may discourage users from trying the product.

In recent years, subscription-based business models have become more common, particularly for content-based platforms like software providers, streaming services, and online news sources. Customers use this model to access information or services by paying a recurring subscription, typically once a month or once a year. The subscription model performs well when companies can regularly offer new material, useful tools, or recurrent services that make the recurring expense worthwhile. Predictable revenue is the main benefit of the subscription model since it enables more efficient planning and scaling for organizations. It can be difficult to keep subscribers over time, so companies must constantly enhance their services to keep churn at bay.

Another popular monetization technique is advertising, especially for websites, blogs, and social media networks that draw a lot of users. Under this strategy, companies provide customers with free content or services in exchange for money from the display of advertisements. The platform owner keeps a percentage of the money advertisers pay to be seen by the platform's users. Since advertisers are likelier to spend on highly-traffic platforms, this model's success hinges on drawing in a sizable and active audience. Additionally, companies must ensure that the advertisements don't detract from the user experience because too much or invasive advertising might irritate users and reduce their interest.

Another revenue-generating method is affiliate marketing, in which companies promote goods or services from third parties in exchange for commissions. Under this strategy, a company or content producer collaborates with an affiliate program to market goods to their target market. The company gets paid when a customer clicks on a link or buys anything after being referred by the affiliate. This tactic works especially well for bloggers, influencers, and websites with a devoted following. Choosing the appropriate items to promote that fit the needs and interests of the target audience is essential to affiliate marketing success.

Effective marketing techniques are crucial for drawing clients, increasing brand recognition, and boosting sales in addition to monetization. One of the most basic marketing techniques is search engine optimization (SEO), which entails improving a website or content to rank higher in search engine results. To increase their online presence, firms must employ SEO, a long-term approach that produces high-quality content, uses pertinent keywords, and develops backlinks. Businesses can improve their overall return on investment (ROI) and decrease their reliance on costly advertising by attracting organic traffic by ranking higher in search results.

Another essential element of contemporary marketing techniques is social media marketing. Businesses may interact with their target audience, increase brand awareness, and increase website traffic by using platforms like Facebook, Instagram, Twitter, and LinkedIn. Social media marketing is the process of producing interesting material that appeals to the audience and promotes interaction, like posts, videos, and tales. To expand their reach and visibility, businesses can target particular demographics with sponsored advertising on these platforms. Companies can use social media marketing to interact directly with customers, create a community around their brand, and get word-of-mouth recommendations.

One of the best marketing strategies is still email marketing, especially for companies with a subscriber base. Businesses may foster relationships, advertise new goods and services, and increase sales by sending customers tailored, targeted emails. Newsletters, special offers, and reminders for abandoned carts are just a few of the uses for email marketing. Reaching customers immediately in their inboxes, where they are more likely to act, is one of the main advantages of email marketing. Companies must ensure their emails are valuable, pertinent, and not unduly promotional to avoid being flagged as spam.

Another tactic that has been popular recently is influencer marketing. This tactic entails collaborating with people with a social media or other platform following. These influencers raise brand awareness and boost sales by using their reach and credibility to promote goods and services to their audience. Since younger, tech-savvy audiences are more likely to trust influencers than traditional advertising, influencer marketing works especially well for companies aiming to reach them. Nonetheless, companies must pick influencers wisely who share their brand values and whose following is similar to their target market.

Content marketing aims to draw in and keep consumers by producing useful, educational, and captivating material. Businesses can position themselves as authorities in their field and gain the audience's trust by creating content like blog entries, podcasts, films, and infographics. Businesses that provide complex goods or services can benefit greatly from content marketing since it enables them to inform clients and solve their problems. Because high-quality material is more likely to be shared and linked to, content marketing can also help SEO by raising a company's visibility.

Another crucial marketing tactic is paid advertising, which includes display and pay-per-click (PPC) adverts. Businesses pay each time a user clicks on a PPC ad, which can be found in search engine results, social media platforms, and other websites. Display advertising, which can increase brand awareness or drive people to a landing page, is displayed on websites and includes banner ads and other visual formats. When the correct demographic is targeted with pertinent messaging, PPC and display advertisements may be quite effective. However, they must be carefully managed to guarantee a favorable return on investment.

To sum up, marketing and monetization tactics are critical to any company's success, especially in the digital sphere. Businesses can profit from their goods and services using monetization techniques like affiliate marketing, direct sales, freemium models, subscriptions, and advertising. At the same time, businesses can increase revenues, brand awareness, and customer attraction by implementing SEO, social media marketing, email marketing, influencer marketing, content marketing, and paid advertising. Understanding the needs of the target market, choosing the best monetization and marketing techniques, and continuously refining these tactics to maintain competitiveness in a market that is changing quickly are the keys to success.

## Maintenance and Future Updates

Future upgrades and maintenance are essential components of the software development life cycle. The success of an application or software system in a constantly evolving technical environment depends not just on its initial development but also on its long-term functionality, security, and performance. Continuous updates, bug fixes, and enhancements are necessary when new requirements, opportunities, and difficulties arise as users engage with the product. Frequent maintenance guarantees that the program will continue to satisfy user needs and adjust to changing specifications. This section will examine the importance of maintenance and upcoming upgrades to ensure long-term software stability, scalability, and relevance, emphasizing the different approaches.

The phrase "maintenance" refers to a wide range of actions to keep a software system functioning, safe, and performing well after its release. Maintenance is a continuous procedure that can continue for the software's life cycle, in contrast to the development phase, which concentrates on creating and deploying a new program. Corrective, adaptive, perfective, and preventive maintenance are the four primary categories of software maintenance. Problems and flaws that appear after the software has been implemented are fixed by corrective maintenance. These difficulties might be anything from performance issues that impair user experience to bugs or glitches that impact functioning. Corrective maintenance is crucial to guarantee that the software functions as intended and that malfunctioning features do not inconvenience users.

Contrarily, adaptive maintenance focuses on altering the software to ensure it continues working with evolving contexts, including new hardware configurations, operating systems, or external dependencies. Software systems must adjust to new platforms and standards as the technological world changes. For instance, a mobile application that performs well on earlier iterations of iOS or Android can require upgrades to continue working on more recent iterations of these operating systems. Similarly, an application that functions effectively on older devices could require upgrades as hardware technology develops to maximize performance on more recent models with distinct processing power or screen resolutions. Adaptive maintenance reduces the likelihood of software obsolescence by keeping it functioning and relevant.

Improving the software's functioning in response to user input, market demands, or technology advancements is known as perfective maintenance. Users may eventually ask for new features or enhancements to current ones. These improvements could be motivated by the desire to link the software with new tools and services, boost productivity, or improve the user experience. Perfective maintenance is crucial to ensure the software keeps up with user demands and maintains its competitiveness in a crowded market. For instance, enhancing the user interface, adding additional integrations, or improving the functionality of specific features can help keep the software's appeal to current users while drawing in new clients.

Preventive maintenance aims to spot possible issues early on and fix them. To make future problems less likely, it seeks to enhance the software's architecture, coding, and general design. Code rewriting, library and framework updates, and adopting improved testing procedures are frequently included in this maintenance. Preventive maintenance increases the software's general stability and scalability while lowering the chance of significant malfunctions. Developers can avoid expensive emergency fixes and guarantee the software's long-term dependability by proactively addressing possible dangers.

Maintenance tasks not only fix urgent problems but also guarantee the software system's security. It is crucial to update the software with patches that guard against new threats as soon as security flaws are found. The world of cybersecurity is always changing, and hackers are always trying to find new ways to take advantage of holes in systems. The software can be subject to attacks compromising user data or interfering with functioning if it is not regularly maintained and security upgrades are not applied on time. The maintenance procedure that guarantees the program stays safe from malevolent actors includes applying security patches regularly, conducting vulnerability assessments, and using best practices for secure development.

Software maintenance is closely related to the software update process. Updates are the main way to provide users with security patches, bug fixes, and new features. A properly run update procedure minimizes downtime while guaranteeing that users take advantage of the most recent enhancements. Updates can be sent in several ways, including automatically and manually. Because they ensure that all users are using the most recent version of the software without needing any action from them, automatic updates are very advantageous. Both desktop and mobile applications frequently employ this strategy, guaranteeing that users get crucial security updates and feature enhancements as soon as they become available.

App marketplaces like the Google Play Store and the Apple App Store usually handle automatic upgrades for mobile applications. Users are alerted when a new app version is available, and the update happens automatically in the background. Some users, however, might turn off automatic updates, which could leave them using out-of-date versions with known problems. This is why designing a system that motivates consumers to update their apps is critical. This could be providing incentives for installing the most recent version or alerting users to important upgrades.

A third-party service or an internal updating mechanism can control automatic updates for desktop applications. Users may occasionally be offered the choice to manually update the program, enabling them to examine release notes and choose when to install the most recent version. Businesses or individuals who wish to manage when updates are applied may find this strategy helpful.

However, it also necessitates greater user participation and may cause delays in implementing critical improvements.

A clear versioning system is essential for monitoring software changes and ensuring users are informed of updates, regardless of the update method. Version numbers make it simple to convey the extent of changes made between releases. Major, minor, and patch are the components of a normal version number. While a minor version update denotes minor enhancements and bug fixes, a major version update denotes big changes, such as new features or breaking changes. A patch update refers to security patches or bug fixes that target certain problems without bringing about significant modifications. Developers may effectively convey the purpose of each update and assist customers in understanding what to anticipate from the most recent release by adhering to a consistent versioning methodology.

Gathering user feedback is crucial for directing the next updates and enhancements and the technical components of changes. Developers can learn how the product is being used, which features are most important, and where enhancements are needed with the help of user feedback. Feedback can be gathered in several methods, such as through surveys, user reviews, support requests, and face-to-face conversations with clients. By looking at this input, developers can rank additions and bug fixes that have the biggest effects on users. Furthermore, user feedback can pinpoint portions of the program that might not work as intended, enabling developers to make the required changes in upcoming releases.

Planning for the software's future in terms of scalability, adaptability, and innovation is just as important as performing routine maintenance and updates to maintain it secure and operational. Software systems must change as new technologies and user needs arise to stay competitive. Future iterations should consider the software's long-term goals, including the possibility of new platforms, features, and connectors. For instance, apps may need to be changed to utilize cloud-based infrastructure, which offers improved scalability and accessibility, as cloud computing continues to gain traction.

The user experience (UX) should also be optimized in future versions. It is crucial to maintain a software interface that is up-to-date, user-friendly, and accessible as new design trends and technologies appear. Frequent UX-improving upgrades can draw in new users and keep hold of current ones. This could entail making the program more responsive and effective, improving the user experience, and providing support for additional devices or screen sizes.

Additionally, the software's compatibility with updated operating systems, frameworks, and third-party libraries should be considered while planning future upgrades. Unupdated software can become incompatible with emerging technologies, resulting in problems with functioning and a bad user experience. Developers may ensure their software stays useful and relevant over time by keeping up with industry developments and remaining compatible with the newest platforms and technologies.

Maintenance and upcoming updates are even more important for maintaining business continuity as more and more firms depend on software to run their operations. To prevent interruptions, software systems must continue to be safe and functional, whether utilized for enterprise resource planning (ERP), customer relationship management (CRM), or any other crucial corporate function. Frequent maintenance reduces downtime, avoids service disruptions, and fixes problems that can impair company performance. Businesses may maintain the dependability and functionality of their software systems, guaranteeing that they continue to satisfy user and consumer demands by updating software and addressing possible hazards before they become more serious.

To sum up, maintenance and upcoming updates are crucial parts of the software development life cycle that guarantee that software will remain secure, useful, and relevant. Regular updates enable software to adapt to user needs, technical improvements, and security requirements, ranging from corrective and adaptive maintenance to perfective and preventative maintenance. Developers can preserve the integrity of the software while offering a dependable and pleasurable user experience by adhering to best practices for upgrades and maintenance. Because technology is changing so quickly, it is essential to keep innovating and adapting through new updates to ensure that software continues to be a useful tool for businesses and individuals.

# CONCLUSION

The fascinating and constantly changing field of "Mobile App Development Basics: Create Apps That Work: Learn App-Building Fundamentals for Android and iOS" presents countless creative opportunities. Throughout this book, we have covered the essential ideas, resources, and methods required to create useful and intuitive mobile applications for iOS and Android. Each chapter has offered an organized process for learning app development, covering everything from comprehending the mobile app ecosystem to creating user-friendly user interfaces, writing code, integrating APIs, maximizing performance, and deploying apps.

An idea and a thorough awareness of the platforms and tools available are the first steps in developing a mobile application. We discussed the frameworks underpinning contemporary app development, the differences between Android and iOS, and how to set up the development environment. We investigated how to develop interesting and usable user interfaces that improve the user experience with a solid foundation in UI and UX concepts. Bringing an app to life requires an understanding of programming fundamentals, and we went over the key ideas needed to create interactive applications.

Developers can produce dynamic, feature-rich apps by integrating apps with external data sources and utilizing APIs. To improve app functionality, we looked at ways to manage networking, safely store data, and include cloud services. In addition to functionality, performance optimization is essential for ensuring an application functions well across various devices. We looked at optimal methods for increasing speed, decreasing memory use, and troubleshooting potential development-related difficulties.

When an app is finished, it takes meticulous planning to launch it successfully. We discussed the procedures for releasing apps on Google Play and the App Store and other revenue-generating monetization techniques. Marketing, user input, and frequent upgrades are crucial for an app to be maintained and improved over time.

Since technology is always changing, developing mobile apps is an ongoing learning process. Developers can produce effective and influential apps by grasping the principles and keeping up with market developments. This book has given you the information and direction to create functional apps, regardless of your experience level.

*Thank you for buying and reading/ listening to our book. If you found this book useful/ helpful please take a few minutes and leave a review on the platform where you purchased our book. Your feedback matters greatly to us.*

www.ingramcontent.com/pod-product-compliance
Lightning Source LLC
LaVergne TN
LVHW022058200926
848481LV00003B/971

* 9 7 9 8 3 4 9 2 7 0 0 2 4 *